CREATION OF A MINORITY GROUP

THE NEW DANGER IN AMERICA'S TRUCKS

© 1996 William B. Trescott

ISBN 1-888748-00-1

Published by Trucking Video
8028 Farm To Market Rd 457
Bay City TX 77414

Printed by Bookcrafters
Chelsea, Michigan

Table Of Contents

Introduction.. 11

I. The Government Is Mine Enemy.................... 19

II. We Are A Minority Group............................ 39

III. We Are Discriminated Against................... 61

IV. Traffic Laws Are Stupid.............................. 91

V. Hypocrisy... 125

VI. The Mechanics Of Safe Driving................ 153

VII. Conclusion... 189

Sample log book page.

Introduction

Introduction

At a deserted inspection station in Nevada, an unarmed trucker is gunned down by a police officer in front of his wife and children. In the worst riot in American history, a trucker is savagely beaten and left for dead for no other reason than that he is a trucker. This is not the playful contest between the "gear-jammers" and "bears" seen in "Smokey and the Bandit" movies. Today, being 2,000 miles from home at the mercy of the local sheriff is no laughing matter. Though unarmed, truckers are regarded as threats the same way racists view other minorities.

Television programs such as 20/20, 60 Minutes, and others portray truckers as womanizing, drug-crazed assassins. "As many as 30 percent of truckers regularly use drugs," says anchorman Tom Brokaw. (The actual federally measured number is two and a half percent). Monster trucks are shown smashing cars in thrill shows, but according to most law enforcement agencies, 80 to 85 percent of car-truck collisions are caused by car drivers, not truckers.

You seldom hear someone say, "I hate Blacks because their

curly hair is ugly and they smell;" or "I don't like Mexicans because their swarthy skin and oily hair makes my flesh crawl." Racism has been replaced with cultural bigotry. Today you hear: "I don't mind Black People living in my neighborhood as long as they act White;" or "I don't mind Jews as long as they go to a Christian church;" or "not all Blacks are niggers — only the ones that don't do what white people tell them!" Minority groups exist in today's media driven society because the majority group is intolerant of personal behavior, not appearance.

Trucks are noisy. In a majority White city, an ordinance can be enacted to divert truck routes through Black neighborhoods to make life better for Whites. By being treated differently than other motorists, Truckers become a minority group. When diverted through a neighborhood of another minority which does not want them, they experience discrimination. In the Los Angeles Riot, the discrimination resulted in truck driver Reginald Denny being nearly beaten to death.

Trucker Ricardo Flores was gunned down in Nevada, another place where truckers are not wanted, because he refused to allow his children to see their father humiliated by laying prostate on the ground.

"Without saying a word, Trooper John Henke fired on the kneeling trucker. The bullet narrowly missed Flores' body and hit his left foot, shattering it.

When her husband was shot, a horrified Laura Flores opened the door of the truck and started screaming, 'No, don't shoot. He didn't do anything!' Henke then aimed the gun at her."
(Sandi Laxson, Landline, Aug.'94, p.14)

Not all laws are intended to be obeyed. Just as a Black Man

cannot change his color when he sees a "Whites only" sign, a truck driver cannot obey the speed limit when weight, height, and width regulations prohibit him from installing adequate brakes. If you go about your daily routine without violating the law, you may consider yourself part of the majority. If politicians enact special interest legislation sanctioning your family's occupation or cultural traditions, then you are not only a minority, but your parents and grandparents are criminals from a family of criminals. If, through the generations, your family has performed a highly technical and dangerous occupation, such as driving trucks, and your forty-ton behemoth cannot tiptoe the delicate ballets and jumping-jacks necessary to obey laws made for cars, you will be labeled a criminal for failing to do the impossible.

Should a democracy have the power to require members of a minority group to report their whereabouts to the government on a 24-hour per day basis? Truckers and other rural people who drive trucks are required by law to fill out log books with 33 textual entries and six arithmetic calculations per day. This means that 231 entries and 42 calculations are required per week; 11,550 entries and 2100 calculations per year; and 346,500 entries and 63,000 calculations in a trucker's 30-year career. If a trucker makes even one tiny mistake, such as misspelling Yreka as "Eureka," when California law says it should be spelled "YREKA," he is a criminal. The odds of a trucker getting all the entries and calculations correct in his lifetime are one in two raised to the 409,500th power! There are not and will never be enough truckers in the universe to insure a single trucker will ever succeed in obeying this law. Because media organizations know that everyone who has ever driven a truck is a criminal, or will soon become one, anyone who drives a truck faces discrimination.

Laws which sanction ordinary human behavior, such as stopping to go to the bathroom, seeking medical treatment, practicing religion, or performing one's occupation have the power to create minority groups where none previously existed. Whether one is a farmer, rancher, or trucker, all rural people depend on trucks for their livelihood. Regulate trucks and all rural people become regulated. A group that previously was in the majority becomes a new minority. *Cowboy, hick, redneck,* and *hillbilly* become the new words of hate just as *nigger, wap, mick,* and *spic* expressed bigotry in the past.

This does not prove that truckers or rural people in general are a minority group — only that they are becoming one. This book does not dwell on past discrimination against existing groups. It explores how minority groups are created by studying a group which is becoming increasingly discriminated against.

*One of the first long haul trucks
rots in the Arizona Desert.*

The Government
Is Mine Enemy

The Government Is Mine Enemy

Truckers used to carry things on their backs. You may think a trucker is someone who drives a truck, but in the 1800's and earlier he was a person who trudged along pushing or pulling a hand truck — not like the teamsters, drovers, and wagoneers who, with mules and oxen, were the real long haul truckers of their day.

In those days, one could not park a team of sweaty draft horses in front of a respectable mercantile establishment. Can you imagine urine, excrement, and flies in front of a ladies' garment shop? Horses had to be fed and watered. Truckers, in flamboyant costumes designed to attract customers, de–livery–ed the goods — that is, removed them from the livery stable where all the drovers, teamsters, and wagoneers went to water their horses. The shipper paid the teamster, but the consignee paid the trucker. If the trucker was nice and well groomed so as not to offend their customers, stacking the boxes where he was told, he might get a tip.

Motorized wagons changed the meaning of the term "trucker." The skills needed to run engines are very different from those needed to drive mule-teams. Few teamsters wanted the earliest motorized wagons because they were unreliable and unsuited to the long haul. The earliest electric trucks had a very limited range. Fuel was not available in most rural Areas. Truckers who delivered freight from livery stables, warehouses, and railroad terminals were the only ones to buy them, so horse-less wagons became known coloquially as "trucks." Truckers became almost superhuman — by tradition still required to hand unload freight and stack it where the customer wants even when it weighs 50,000 pounds!

The earliest truckers in America were slaves. A courtesan of Queen Elizabeth the First named Sir Walter Raleigh wanted to get into the drug trade and sell tobacco. Native American Indian tobacco was a lot stronger then than it is now. Modern cigarette tobacco has been bred for productivity, not potency. There was money to be made if this could be grown on a large scale. The Queen had a problem with Catholic Scots who were fleeing from Scotland to Ireland (following the execution of the Catholic Mary Stuart, Queen of Scots) and stirring up disloyalty to the British Government. These Scotch-Irish as they came to be known were probably truckers even before they came to America, since there were no other jobs available in Ireland for refugees. Elizabeth ruled that all people born in Scotland were English subjects. Because Catholicism was against the law, she felt they could legally be captured and sentenced to whatever punishment she desired — such as being sent to Virginia as slaves to work on Raleigh's tobacco plantations.

These Scots did not last very long as slaves. They escaped — to coin the expression, "Scot-free." It had been believed that the Cherokees would kill any white they encountered because well—armed soldiers were always killed. The Cherokees did not kill people at random — only those who threatened them. Helpless escapees were welcomed as members of the tribe. They taught the Cherokees new technologies such as wheels, gunpowder, and how to make muskets. They helped the great Chief Sequoia invent a written language for his people. The Cherokees became the first "Civilized" tribe.

The English provided these Scotch-Irish escapees with an income by imposing import and export tariffs. The Boston tea party, which sparked the American revolution, was a protest over tariffs. Every time goods crossed from one colony to another (such as from Virginia to Massachusetts) a tax had to be paid.* The British Army enforced the tax on land and the Royal Navy collected it on water. Anyone who could ship goods without being taxed stood to make a hefty profit. Those who used the water route were called pirates and those on land: "truckers."

For obvious reasons, large horse-drawn wagons could not be used. There were few roads other than Indian paths. The ruts and tracks horses create would attract the attention of the British army. Horses were rare in the colonies at first. Owning a Clydesdale to pull one's wagon would be like having a Caterpillar 3408 V-8 in the doghouse today. Mountaineers would truck or carry the goods up through the gaps or passes were they could be loaded onto sledges for a precarious slide downhill to the back-roads and rivers of Kentucky, Tennessee and West Virginia. They could then be shipped without harassment from the government. If you have a hard time believing that truckers traveled long distances pulling carts by hand, go to the temple in Salt Lake City. The only statue is of a trucker and his family. Those who live history remember it. *(see photo page 23)*

* Trade between colonies was illegal unless loaded onto a British ship.

Truckers were smugglers in the eyes of the British government and suffered the worst form of persecution. Few Scotch-Irish women were imported as slaves, so marriage to European women was rare. The army slaughtered Cherokees and other Indians on sight, leaving whole tribes of women and children without men to protect them. Out of marriage of convenience and necessity, a new race was born: the Hillbillies.

The term "Hillbilly" is loosely applied today to any rural person. Though few would like to admit it, the vast majority of rural people have at least some Indian blood. "Hillbilly" is a racist term every bit as virulent as nigger, spic, wap, kike. It denotes mixed ancestry. Rural people are also called rednecks; though this has more to do with the red scarves they wear to keep the dust out than the color of their skin. People who are half–Black look Black and are accepted in the Black community as Black. People half–Oriental look to Europeans to be Oriental, but in Japan and Vietnam they are discriminated against because they look White. People half–Native American look White except for a few telltale signs, unnoticeable to us but excessively important to puritans who devoted themselves fastidiously to Biblical injunctions against miscegenation. The red man was heathen and Catholics were heretics. Hillbillies, being both Catholic and Indian, were *heretical heathens*.

Hillbillies had several beneficial effects on this country. First, they were the first people who were really and truly "American" — born on this continent with no European ties and prejudices. They combined the technology of Europe with the medicinal and agricultural knowledge of the Indians. Europeans also had medicinal knowledge, but it applied only to European plants. Indians knew how to use American plants to cure illness — or cause it. Many Hillbilly women were tried by the government for witchcraft. You can see the racism today in our children's

The Emigrant Monument; Tempe Hill, Salt Lake City, Utah

The Emigrant Monument; Tempe Hill, Salt Lake City, Utah

The Emigrant Monument; Tempe Hill, Salt Lake City, Utah

Halloween costumes; the ruddy skin, hooked noses, straight black, or wiry gray hair as Native Americans have when they get older, and the pointed Norman hat and cloak, which was the height of fashion for widows at the time. We all know what witches are supposed to look like. The Puritans did not understand herbal medicine, but they could see its effects. If there was a Puritan family whom an old Hillbilly woman disliked, she could poison their well. Indians and Hillbillies did not need wells.

The mixing of races gave rise to a new religion known today as Pentacostalism. This began as a mix of Christianity with the spiritualist and animist religions of the Indians. Catholicism was illegal in the colonies. When they lost their priests, the Catholics felt spiritually lost. They had to find their own spirituality. This came in the effect of adopting native American ritual. Quakers and Shakers would whirl around doing Indian dances — a far removal from the subdued liturgy of the Catholics and Anglicans. Pentacostals would wave their hands in the air in religious ecstasy, receiving spiritual gifts of prophesy and talking in tongues. Snakes are still used in some Christian worship services and ordeals such as drinking poison or snakebite are believed to demonstrate God's favor — just as in Native American religions. The government burned many people at the stake just for being Quakers.

The most important gift the Indians bestowed on us was democracy. Indian tribes have always been democratic. Europeans always believed in the divine right of kings. In the Iroquois confederacy all chiefs were equal regardless of the size of their tribe. Hillbillies believed that all men were equal so long as they had a gun. The belief that guns equal freedom led to the revolution. It also gets truckers into trouble today because carrying guns in trucks is illegal even though about a third (two thirds in some companies) still do.

(At the time I am writing this a self-styled rural Messiah named David Koresh is fighting tanks near Waco, Texas in defense of his bizarre religion.)

The revolution would probably have been unsuccessful if it were not for the secret trade routes of the truckers. Surveyors like George Washington did more than measure plots of land. They were also spies for the government. George must have met many a trucker who would later help the Minute-Men vanish into the woods, only to reappear and attack the British somewhere else. His army would have been unable to fight without the truckers' discrete logistical support.

European combat strategy was based on use of the huge Brown Bess musket which, with its gaping .75 (three-quarter inch) caliber smooth bore, was much more a personal howitzer than a rifle. Combat was fought at distances of more than a quarter mile or half kilometer, with the soldiers huddled together, shoulder to shoulder, to reduce the effects of each other's muzzle blasts on their ears. Each soldier would fire in turn, adjusting his sights according to where the last man's shot landed. The wind, more than aim, had the greatest effect on accuracy. Once the enemy was zeroed in, they would be mercilessly slaughtered.

A three-quarter inch ball was big enough to be lethal if it were merely thrown at someone's head. Launched out of a six foot long gun at a thousand miles an hour, it could shatter stone walls and knock down trees, creating flying debris more lethal than the balls themselves. The enemy could not run or hide or surrender. Soldiers could only stare at each other across an open battlefield and hope to shoot the foe before being shot. They took cover only in the huge volumes of smoke the primitive black gunpowder produced, thick enough to conceal a quick-march parade left or parade right as phalanxes of men moved around the open battlefield like chesspieces. The Brown Bess musket made the British Army the most formidable in the world.

To combat this weapon, Americans resorted to guerrilla terrorist tactics, ambushing the enemy with highly accurate hunting rifles at close range where the caliber of the gun did not matter. The Brown Bess used huge amounts of powder — 200 grains per shot comparedto the 70 to 100 grains a hunting rifle used. The .75 caliber lead shot weighed three times as much — the weight cutting down on the soldier's mobility. The Americans could outmaneuver and harass the British until they ran out of ammunition.

These sudden surprise attacks would not have been possible without the effective concealment of American military movements and logistical supply. The British resorted to quartering troops in private homes to keep them from being shot as they slept.

After the revolution, truckers found themselves ostracized as much by the new government as the old. Roads, secret or not, have to be maintained. This was done by highwaymen who would rob (collect tolls from) passers-by in order to pay for highway maintenance. Naturally, truckers and poor people would get a discount and people they did not like (aristocrats) would be charged a prohibitively high fee, just as truckers pay higher tolls than cars do today. Once tariffs between colonies were gone, there was no reason to keep trade routes secret; so the rich started to complain about being soaked and campaigned for big government to take over privately held toll roads or turnpikes (named for the rotating barrier used to force people to pay toll.) Imagine the shock farmers would have felt if their land was collectivized or if ranchers were told their cattle did not belong to them anymore. The truckers' economy was destroyed and a precedent was set that allows any stupid idiot, qualified or not, to drive an unsafe type of automobile four times as dangerous as a gun. Of the sixty or so corporations that existed at the time of the revolution, more than half were private toll roads.

The Brown Bess musket.

A typical wagon train in 1868. Note that the drivers preferred to ride nice, smooth mules rather than cargo–filled wagons. There are no women in the picture.

Wagons full of settlers were no more common than household moving vans are today. Photo credit: Colorado State Historical Society

The wagon in the background proves that Apaches were more technically advanced than most historians give them credit for.

Photo credit: Aultman Collection, El Paso Library.

27

In 1794 truckers revolted, almost dividing the country in two, when treasury secretary Alexander Hamilton decided to impose a tax on whiskey. Beer, especially dark beer in those days, was not merely a beverage, but a food. The alcohol in it allowed it to be preserved through the winter or transported long distances. Fermenting malt was less laborious than grinding grain into flour and the resulting beverage was almost as nutritious as bread, as any overweight beer drinker can attest. Drinking six droughts a day would make one just as fat as eating six loafs of bread a day. Distilling beer produced whiskey, which could be used to purify water — an essential product for both whites and Indians in a land without doctors or medicine for treating waterborne diseases such as Plague or Cholera.

The importation of Africans to replace the Scotch-Irish as slaves dealt the Hillbilly community a double blow. If farm labor was essentially free, why would any planter hire a man to do it? Since the best farmland was already taken by the wealthy, Hillbillies had no choice but to move west into the mountains of the Carolinas, Virginia, and then into Kentucky. They pushed the Cherokees, who had declared their independence from Britain at the same time as the Americans did, all the way to Oklahoma and the Dakotas out of Minnesota into Crow territory, who in turn pushed the Crows into Montana. The Alogonquin and Iroquois were pushed out of New York into Canada where they had committed genocide against the Hurons. The Indians did not die of smallpox as the history books say. They starved to death when they were pushed off their land. One of Abraham Lincoln's first acts as president was to order the largest mass execution in American history. A group of Dakotas had dared to fight whites in defense of their land. The whites wanted to commit genocide against the whole tribe. Lincoln wrote in his own hand which of the Indian soldiers taken prisoner would be made an example of and which

Though clothing styles may be different, these Apaches from White Mountain are as well-groomed, fed, dressed and well–to–do as a typical White family of the period. They were the only Arizona Apaches who did not lose their land. Only those Indians sent to reservations lived in poverty.

Many truckers continue to value their heritage.

29

would be allowed to live. There was no Geneva Convention in those days.

On the whole, Hillbillies got along with the Indians very well. They traded furs for whiskey, which was necessary to protect them from plague, cholera, and the white man's other waterborne diseases. Mountain men married and lived among the Indians. Spain and France were enemies of England. Before the revolution, trade with these countries was illegal. Truckers were able to make money importing wine and dried peppers from greater Mexico and Acadiana, now called Louisiana. After the revolution, trade with Spain and France became legal and trade with England impossible because the English had a boycott against the United States. There was money to be made in trading with Canada.

Trade allowed Indians to become quite civilized and some even quite wealthy. Sequoia, Chief of the Cherokees, wrote a constitution used by his people to this day. Navajos raised sheep and mined silver. Their blankets were considered the best in the world. Before Indian blankets, woolens came from England, Scotland, and Ireland. You could have any color and style you wanted so long as it was gray. Not only were Indian blankets more colorful, the natural dyes did not fade or destroy the fabric. Blankets made during this period have become collectors items worth thousands of dollars. At one time nearly all American woolens — blankets, socks, and nightcaps were made by Indians.

Gray flannel suits are still popular among whites, but Indians, Hillbillies, and today's truckers prefer to dress flamboyantly: blue jeans, plaid shirt, red neckerchief, white hat, and elaborately tooled leather belt and boots became their trademarks. Silver belt buckles and bolo ties the size of your hand are worn by the well–to–do. Drivers' boots can be distinguished from cowboy boots by the metal point on the toe that would be gouged into the animals' rumps. They did not wear spurs. The best animals were always put in front with the lazy ones behind where they could be encouraged.

Though public school history books give Lewis and Clark the credit for being the first whites to cross the continent, they had to go to Oregon instead of California because the Spanish beat them by nearly a century. Many wagons made the trek yearly. When gold was discovered in California in 1849, underpaid sailors jumped ship to pan for gold, leaving wagon trains the only way of shipping freight across the continent. Wagoneers made almost as much money as the miners did, keeping them supplied with clothing, sugar, nails, whale oil, whiskey (which was essential to treat water), and any other goods they desired. Gold was cheap and prices were exorbitant, but nobody cared as long as there was plenty to go around.

Drivers made a killing. They would haul trade goods from Denver to Truckee along the Oregon trail, exchange their loads for Haida baskets and gold, and return by way of the Nevada trail, were they would then load up with silver, blankets, and other woolens and return with to Denver by way of the Santa-Fe trail to wait out the winter. The money they deposited in western banks later allowed them to emigrate.

Although public school history books say that the west was settled by pioneers, the only groups that can authentically be called pioneers were the Mormons, Quakers, Hutterites, and bizarre cults that fled religious persecution. Hillbillies did not go west before the settlers — they were the settlers! Just a few trips to California and back was all it took to earn enough money to start a ranch or farm.

Cowboys saved their money to buy saddles so they could become buckaroos (an Anglicization of the Spanish word vaqueros). They delivered horses to market on the hoof in those days, same as beef. A horse that arrived saddle–broke was worth more. A cowboy who could afford his own saddle could become a drover and do the difficult job of rounding up cattle on the open range — earning twice as much as a mere ranch-hand. It was an unwritten rule that a cowboy could not marry unless he had a rig to take his bride home in. You could always bust sod to build a house, but you had to have the rig to take the crops to market. Marriages and new trucks still seem to go together though the government now makes it illegal for many truckers to take their brides along. Perhaps only one in five wagons in a train had a family on board. Those that did were owned by men who had made the trip many times before and knew where they were going.

The one thing the Indians wanted more than anything else was guns and ammunition. Game became wary and hard to find. Aside from their survival value in hunting, guns allowed the Indians to defend their territory against the whites. This was not a problem for Hillbillies, now known as Rednecks, because they knew native-American customs, being half Indian themselves, and were able to get along. Most Indians had an animist religion in which they believed one could adopt the animal spirits of the clothing one wore. By wearing clothing made from docile animals such as deerskin or a coon skin cap, traders could proceed across Indian territory unmolested. The Indians enthusiastically wanted to trade. A few cheap gifts was all the toll one needed to pay to cross Indian territory unmolested. Those who tried to shoot their way across usually got shot themselves.

The government wanted to settle Oregon before the British built up enough of a fleet in the Pacific to impose their sovereignty over the region. Wildly exaggerated claims made about the quality of Oregon soil and the ease of getting there lured stooges out of overcrowded eastern cities to settle. These people knew nothing of Indian custom. They wore pretentious clothing as though they were trying to out-chief the chief, and did not bring any gifts. They were seen as a threat to Indian sovereignty and were killed. It pays to know the language of the people one does business with.

The government blamed Hillbillies or Rednecks for the constant Indian wars. If the Indians had not been given guns they would not be able to fight. Drovers were seen as co-conspirators, aiding and abetting the crimes. To be an Indian lover was almost as bad as being an Indian. Besides, escaped slaves were mixing the pot. Indians and Quakers were the only groups that would take ex-slaves in. The Underground Railroad was virtually run by Quakers. At the same time that whites were being killed, the west was being populated by so called "inferior races."

When I was young and unemployed I had the idea to ride my bicycle across the United States. One of the more unusual towns I came to was Vandalia, Michigan. This was formerly a Quaker settlement that is now inhabited by the blackest, darkest skinned Blacks I have seen. Descended from escaped slaves and almost purely African, everyone I saw seemed to give me a fierce look, as though I were unwelcome. These people own their own businesses and are not dependent on whites. They don't have to be nice. They can treat Whites like Whites treat them. An Indian tribe in Connecticut got rich when it opened a gambling casino. Whites tried to close the casino by challenging in court whether they really were a tribe — because their Chief looked almost Black!

The government did all it could to disenfranchise the original inhabitants. Whenever the army marched in, property lines and title deeds were ignored. Daniel Boone lost his land in Missouri three times: first when the Spanish marched in, then with the French, and finally with the Yankees. There is no doubt which side Hillbillies were on in the War Between the States. The Stars and Bars can be seen on many a sleeper curtain and radiator grill to this day. In actual fact, drivers gave logistical support to both sides. The Yankee army had a habit of commandeering wagons. "Either you haul for us or we'll take your rig and try you for treason." Some drivers fought against the South because the South represented European aristocracy — which they opposed.

The War Between The States was not about slavery until two years after it started.* Slavery was ending because technology such as the steam engine, cotton gin, and McCormick's reaper was cheaper than owning slaves. The government did not want bands of freed slaves settling the west, so they invaded the south to institute land reform, hoping Blacks would stay put if they owned land. In the end, it turned out to be a big grab for power by northern carpet baggers to displace the original southern inhabitants.

In the Oklahoma land boom, Yankees were told they could just walk in and take land from the Indians. The seven civilized tribes who lived there, including the Cherokees (called "Sooners"), considered themselves to be an independent nation. The Boomers were shot or simply ignored. The few who stayed had to settle on dry plots of land that turned into a dust bowl during the depression of the 1930's.

When the American Army invaded Utah, hundreds of young Mormans took up arms and pledged to fight to the death. The President of the Mormans, along with the Ute tribe, considered Utah to be an independent nation. To avoid the burning of Salt Lake City the same way Atlanta was burned in Sherman's march through Georgia, the President agreed by divine revelation from God to change the Morman religion and ban polygamy. Many Mormans fled to northern Arizona to live among the Indians. Polygamous families can be found there to this day.

Many years prior, Sam Houston, founder of Texas and a former governor of Tennessee, showed up in congress wearing

---

* In fact, the Czar of Russia freed his slaves one year before Abraham Lincoln did. The British Empire made slavery illegal in it's territories seven years prior.

34

Indian garb to protest broken treaties and ill treatment of Indians and other minority groups. He wisely saw that the government could not be controlled and that going to war was fruitless. He tried to reach accommodation with the Yankees. By bringing Texas into the union voluntarily, he avoided our being conquered. Today, Texas still controls most of its land.

Truckers have a unique perspective on history because we see it firsthand. I visit every major city in the United States a least once a year. I talk to people who were there. Whom should I believe: a person whose grandfather set them on their knee and told them the way it really happened; or public school teachers, fresh out of college, who's minds have been programmed by propaganda from government–approved history books? To be sure, not everyone tells the truth and most personal histories are exaggerated and vainglorious, but they are not less accurate than government history, just different.

Hispanic citizens with brown skin must show their papers to freely travel within the country.

We Are A
Minority Group

We Are A Minority Group

Hitler killed Gypsies before he killed Jews. According to Simon Weisenthal, the famed authority on Naziism, a greater proportion of European Gypsies were killed in Nazi death-camps than Jews. While the media likes to portray female Gypsies as palm readers and fortune tellers, their men were the truckers of Europe and traveled everywhere from Spain to India gathering, like American Indians, a medicinal knowledge beyond that of most Europeans. They too were historically tried for witchcraft and burned at the stake.

It is not known how many Gypsies emigrated to the United States. They were classified as Hungarian, German, Polish, Slavic, and even English. Their influence on America can be read on the back page of every newspaper — their religion was Astrology. Decorations on wagons, hats, vests, and boots are so similar between American and European truckers that many drivers today must have been descended from Gypsies. Some older friends of mine remember visiting Gypsy camps when they were growing up.

The term "Gypsy" originally applied only to Jews. It referred to the Jews having come out of Egypt as mentioned in the Bible. The term came to apply to modern Gypsies, who actually come from northern India, when discrimination against Jews prevented Jews from traveling and modern Gypsies took over their trade routes. The term "Gypsy" can therefore rightfully be applied to any trader or trucker in the transportation industry.

The Jews were the original truckers of Europe. They helped form an organization of trading families known as the Hanseatic League which invented many of the organizational structures vital to the functioning of the modern trucking industry. The most important was the hub–and–spoke system of freight distribution. Counts, Barons, Dukes, and other aristocrats extorted tolls for passing through their territories. To minimize these expenses, these Jewish families shipped goods by water on an indirect route to the island of Wisby (Gotland) in the Baltic sea where they would be unloaded and put aboard other ships with other goods bound for the same destination. Because the goods did not go overland, no tolls were paid and they made extraordinary profits. They were also regarded as smugglers.

Another great innovation of European Jews was the use of cut diamonds as a common currency. Gold coins were used as money since Roman times, but it was easy to trim or file the coin or to alloy the coin with a base metal like lead. The Jews had a system of appraising the value of diamonds so that a diamond in Poland would be equally valuable as one in Spain. Diamonds were easily hidden and even swallowed in the event of theft. Whole families were often butchered because Europeans believed an old folk wisdom that diamonds came from the bowels of Jews. Aristocrats would often detain these truckers on trivial offenses until the gems passed through their systems just as modern courts extort money from today's truckers by threatening to delay them. Nevertheless, such innovations gave these Jews enormous control over European trade.

Hitler did not want any minority group to have such power. He considered Gypsies a threat because of their involvement in the drug trade. Their religion, related to the Parsis of northern India, involved the use of opium. In America, tobacco laced with opium allowed Black slaves to feel no pain when they worked — making plantation owners rich, as professional athletes are injected by their owners with Novocaine today. Addiction made escape from slavery impossible because of sometimes fatal withdrawal symptoms. Due to cultural tradition, generations of Blacks are addicted to heroin to this day.

Hitler feared German youth might be seduced to use opium because of the hardships created by the world–wide depression of the 1930's. Though only a tiny percentage of Jews were involved in organized crime, he scapegoated the entire group — all six million of them. Then, as now, drugs were smuggled. The only way to stop the drug trade is to control all trade. Hitler killed Gypsies because, like the Jews, they had gained too much economic power. Not only were German teen–agers addicted to their drugs, but they imported a myriad of other products that Germans had come to depend on. Europe was embroiled in a trade war. Germany raised tariffs to pay for reparations after the First World War. In retaliation, the French, Czech, Polish, and others raised their tariffs. Before the income tax, tariffs were the primary way that governments raised money. The trade war set off a depression that created a huge demand for social and welfare spending. The smuggling of goods created unemployment in domestic manufacturing as well as a loss of government revenue. Tariffs discouraged importation of goods, helping to prevent a loss of jobs and pay for welfare spending at the same time.

As trade decreased due to tariffs and border enforcement, Jews and Gypsies, as well as German truckers found themselves increasingly unemployed. Hitler did not want Germans to be unemployed when people of so called "inferior races" kept their jobs. Germany and other countries like Poland required literacy tests in order to obtain a driver's license. The Jews and Gypsies spoke Yiddish, not German, so they had to take crash courses in German if they wanted to keep family trucking businesses handed down for generations. Most flunked the test. Once they were unemployed, they were economically helpless and the Germans could do what they wanted with them, such as putting them in cattle cars and sending them to death camps.

I remember when I took my first federally required Commercial Drivers License test. I scored 98% even though I deliberately answered a third of the questions incorrectly. The only question I missed was answered correctly, but the officer administering the test, who had never driven a truck in his life, put a blue stencil over my answer sheet and said I was incorrect. The only reason I passed the test at all was that I had studied the Uniform Model Commercial Motor Vehicle Handbook and highlighted those incorrect portions in red. I studied these especially so I would not be caught by trick questions.

A Hispanic man taking the test with with me had ten years experience. He flunked on his second attempt with the same score he got the first time — 30% wrong and was fired from his job. They offered to let him take the test in Spanish, but since his family had always lived in Texas even before there was a Texas, he couldn't read Spanish. Public schools taught him to read English well enough to take the test but not well enough to read the manual to find out the answers to the trick questions. I told him he was lucky America doesn't have concentration camps. He said he didn't know what I was talking about, but I'm sure he knew how the Jews and Gypsies felt.

The reasons for discrimination have historically been

This nice officer wrote me a fine after I was rear-ended by a State of Virginia highway construction vehicle.

Though I was found innocent in court, I was forced to pay $273 to repair the dent in his bumper.

43

economic. Deferential laws exist to favor one group over another — such as Germans over Jews and Gypsies, or European Whites over Indians and Hillbillies. It makes no sense to remove an experienced driver from the road to be replaced by a novice who has a five–times higher likelihood of killing someone. It doesn't make sense to deny a Black man the right to use the toilet when the Ku Klux Klan guy who thought he was getting cleaner rest-rooms tracks human feces from the woods behind his house across his living room parlor.

What is in the mind of a police officer when he steps up to a Black man 50 years his senior and calls him boy? What made the Chinese Red Guard arrest college professors to make them work in the rice paddies — causing a "lost generation" of youth who will never read and write? What made the Nazi Brown-shirts burn the temples and synagogues, then beat a Catholic priest who ran in to save the Torah? What is in the mind of the trooper who pulls over a trucker for a speeding infraction when he knows the trucker is a ten–times better driver than he is?

I was rolling down from "chain–your–cow–up" (Chiracahua) pass in southern California a few years ago when I saw a highway patrol officer chasing a car about a mile behind me. I blocked the car and the cop caught him immediately. Instead of pulling over the speeder, he signaled me to pull over. I don't have an engine brake and my brakes were hot, so I made him chase me to the bottom. When I finally managed to stop, a kid in a brown shirt got out of the car. I fanned the pedal as he walked up. Instead of offering to help chock the wheels, he told me to get out of the truck! I know he was nuts, but I didn't want a ticket so I warned him I had hot brakes, then did what he said. I figured I could always reach in and hold the brake pedal down if the truck started to roll. Besides, the cop car was parked directly behind me; it would not have rolled far.

44

The kid was out of his mind. He insisted on examining my paperwork before chocking the wheels! I reminded him I had hot brakes and reluctantly gave it to him. He thought I was being disrespectful because I kept looking at the front tire to see if it was moving. He said he was writing me a citation for an illegal lane change maneuver, cutting off a car. I told him I was sorry and that I only did it because I thought the car was driving recklessly, resisting arrest. The car had indeed been driving fast, but not as recklessly as the cop was driving. He said I had messed up his VASCAR reading — a type of stopwatch method for catching speeders in which elapsed time between mile posts is calculated to determine speed.

I again reminded him I had hot brakes. The officer hissed and gave me the stupid despotic pout teen-agers get when upset.

"Why do you keep on mentioning that," he asked, "this business about the brakes being hot?"

"The parking brakes aren't set."

"You know I could give you a hell of a fine for that!"

"What," I asked, "and warp the drums; what good would that do?"

Just then he glanced back to see that his car was parked directly behind my truck. He shoved my papers into my chest, and ran. He sped away without explanation. The truck never moved an inch. Imagine what hospitals will be like when politicians start telling doctors how to practice medicine.

People say that Hitler was some kind of monster, maybe a psychopath. I think he was just a shrewd politician. Politicians need to be as cold as fish and slimy as butter to get elected. To find out what kind of man Hitler really was, lets compare him to the present president at the time I am writing this: Bill Clinton.

CREATION OF A MINORITY GROUP

Hitler did drugs...
  Clinton did too.
Hitler dodged the draft...
  Clinton did too.
Hitler womanized.
  Clinton did too.
Hitler never in his life did an honest day's work.
  Clinton didn't either.

The only difference between the backgrounds of Hitler and Clinton is that Hitler eventually did join the army and distinguish himself on the battlefield when his entire unit was massacred. Perhaps seeing all his buddies killed did turn Hitler into a monster, but Roosevelt also had concentration camps. American citizens whose grandparents were Japanese, were interned at Manzanar, Gila River, Tooele, and other places. I've been to Manzanar. Unlike Auschwitz, the American government wanted to forget. Everything has been raked clean except for a pagoda–style security shack that still stands like some kind of shrine. The reason nobody died there is because nobody bombed the rail lines needed to bring food into the camps. Most Jews and Gypsies died of malnutrition and disease due to food shortages, not gas chambers.

Our ally in Russia, Uncle Joseph Stalin, killed more people than Hitler did. Roosevelt warmly shook his hand and kissed his cheeks — the man who two years afterward started aiming atom bombs at our children. People say that the Germans are naturally evil for starting so many wars. There are plenty of politicians in the American government who have done the same thing. That is how the Vietnam war got started. That is why tanks destroyed a religious cult in Waco, Texas rather than have a deputy sheriff simply arrest the cult leader when he came out for his daily health run.

All this begs the question: "What is a minority group?" Is a Black millionaire underprivileged? Africans and West Indians are blacker than most Afro-Americans but their average standard of living is slightly better than Whites. Oriental immigrants have an above average standard of living. Some say Homosexuals are a minority even though, not having wives and children, their standard of living is way above average. Blacks are discriminated against because they are descended from slaves — not because of their skin color. Hillbillies in West Virginia are also descended from slaves and they are also equally poor.

A Black millionaire who cannot join a club frequented by White millionaires might seem to be a victim of discrimination, but truck drivers are not allowed to join such clubs either; so I am also underprivileged. A Black is not a minority when he is turned down by a club when the majority are also turned away. Discrimination in this sense is occupational; the Black man is turned down because he is descended from slaves and I am turned down because I am descended from truckers. Children of millionaires are not turned down by such clubs even though they personally have less money than I do.

The only definition of minority that holds up to examination is the legal one: A minority is a group of people oppressed by particular laws which are irrelevant or inconsequential to the majority group that enacts the laws. By this definition a Black millionaire would be considered a minority if he could not sit in the front of the bus or use a public toilet. A homosexual would also, if it were illegal for him to dance with his partner at a party when heterosexuals are allowed to dance.

The problem with the legal definition is that most minority group membership is voluntary. A homosexual can dance with a woman if he chooses. A Black millionaire can hire a limousine instead of riding the bus or have plastic surgery to change his features. A Jew can become baptized. You are only a minority if you are unable to conform to the standards of the majority such as in the case of discrimination due to your parents' or grandparents' occupation.

Truckers are not considered a minority under existing law because occupations are voluntary. If you had several job offers when you decided to become a trucker then you have no right to complain if you are harassed by the government. When I became a trucker, I had only one job offer. I have had no other offers since. I must drive or I cannot support myself. The United Nations has declared that the right to work and support one's family is a human right. My minority status is not voluntary.

In time, I could be retrained to do something else — like write this book. Minority group status is relative to the ease with which you can change. A homosexual can stop having sex in an instant. A Jew can become baptized in about two minutes. A Black can have plastic surgery to look White in about two months. A woman can have a sex change operation and become a man in two years. A truck driver can go to college and get an equivalent job outside the industry in about four years. If it takes you longer to get another job that will support your family than it takes a Black man to have plastic surgery to become White, then your occupation makes you a minority. If your father and grandfather were truckers and trucking is the only skill you have ever had and you have a family to support, preventing you from taking time off to go to college in order to get a better job, then you are a minority and you are entitled to minority group protection.

I once met a man who said that in ten years in the restaurant business he has never had a White Southerner apply for a job. He said he doesn't hire Blacks because they are bad for business. He said he can tell a person is Black just by listening to their accent

Some of the stickers needed to drive a truck legally in the lower 48 states. Fines for not having one of them range from $100 to $1,000.

over the phone and he usually discourages them from coming in for an interview.

"You know," I said, "Black and White Southerners talk pretty much alike; are you sure that no Southerners ever called you?"

The main reason for discriminatory laws is to generate income for the majority. The Ku Klux Klan knows perfectly well that Blacks won't defecate on the ground because that would be beneath human dignity. The bathroom laws were designed to sap money out of the Black community. Little white boys used to charge a dime to guard the toilet while a black man used it. Hitler used eugenics theory to justify taking jobs away from Jews. The cops know perfectly well that when truckers overheat brakes to obey speed limit laws people get killed; but they continue to enforce the laws in order to generate revenue from speeding fines.

Suppose that for a Black man to get a job he had to have three licenses, 70 stickers, and 200 registrations all of which required 200 hours of paperwork and 200 pieces of mail per year and a White person could get a job with one license that took fifteen minutes to get. Rural people in the United States have the choice of only a three main occupations. These are farming, ranching, and trucking. Most of these require the ownership of some land. All of them require the driving of trucks. Rural people cannot feed their families without driving trucks. They must go through this excessive licensing procedure just to make their trucks legal to drive. A city person can get their car registered and go find a job in fifteen minutes.

I once met a man who was four generations in the candy business. He made adorably edible soft candy-cane baskets full of little soft candy-canes. I bought one for my mother at the wholesale price; but could not resist the temptation of it sitting on the dash and ate it. Candy making is not exactly a macho occupation. The government humiliated him by giving him huge fines every time he tried to truck his product to market, so he turned to a life of crime. He now delivers his product in a rented U-haul trailer attached to his unmarked personal van — just like a drug dealer. He was worried that the government might notice his frequent trips as his van was dangerously overloaded. He made me swear not to tell. He was worried that he might kill someone with his overloaded vehicle but the government won't let him drive a truck because licensing procedures are too complex and expensive.

Many laws that seem to be discriminatory against truckers are really aimed at rural people. Truck farming, where a farmer uses a truck to take his goods to market himself to avoid the middleman, is the only type of farming that is really profitable for small family farms today. City folks do not like to see farmers get rich because with wealth comes political power. Cities tend to view rural areas the way the Japanese viewed the South Pacific before they bombed Pearl Harbor. Japan is like a giant city. They invaded and conquered to take the resources they needed to build a modern industrial economy. They bombed the American fleet because they thought we Americans were going to stop them from industrializing.

Cities in the United States are now doing a Pearl Harbor on rural America. They regulate farmers, ranchers, loggers, and truckers to make them dependent on wealthy city middlemen — rich beyond our dreams. Election campaigns are so expensive these days that politicians have to associate with millionaire contributors if they want to be elected. They are afraid that if we rural people controlled trucking, we might go on strike and deprive millionaires of resources they need such as food, gas, and garbage disposal. Such a strike would eliminate rural poverty and deprive politicians of their millions, shifting political power out of the cities.

I've been told that the difference between a big truck and a monster truck is that a monster truck has a Black man behind the wheel. Racists become afraid that when they see an "inferior" in a position of power over them. They think Black truckers will take revenge for all the discrimination they suffer by smashing their BMWs.

This raises the question: "Why do car drivers drive near big trucks?" I've seen car drivers drive beside me for several minutes at a time, speeding up if I speed up and slowing down if I slow down. They weave out of their lane and force me onto the shoulder. They seem to like to drive near a vulnerable place like the fuel tank, never by the trailer. Friends of mine have been run off the road by such cars.

No trucker would deliberately run off the road if it were not for the treatment they receive from cops. Illegal cab searches, drug tests, logbook inspections, and detentions, all conspire to make truckers paranoid about getting into collisions. These laws are not just discriminating, they are intimidating. The safest thing a driver can do in a reckless driving situation is to grip the steering wheel tighter. Destroying your truck to avoid reckless drivers is stupid! A friend of mine broke his wrist when his steer tire was hit by a car. He did not lose control of his truck. Losing control is far more dangerous than hitting something. The main cause of death among truckers is roll-over accidents, not collisions.

In a car race, drivers demonstrate superiority over each other by passing. In race relations, a White expresses dominance over a Black man by calling him "nigger." People play sports and support sports teams to demonstrate their superiority over others. When the Houston Rockets beat the New York Knicks, we said, "We won and they lost." "We Texans are better than they are." Actually, both groups of fans lost. The owners and players won and fans are out the price of the tickets. The reason people pay money to support sports teams is that in a larger sense, a sports victory proves the superiority of one city over another. If Houston is doing better economically than New York, then Houston fans will have more money to spend on sports tickets. The team owners will then have more money to hire better players, who will win more games, and prove superiority. In Europe, supporters often spend a lot of money on sports even when their town is not doing well economically. Football matches often become violent because the teams represent racial or ethnic groups rather than mere cities. A Black basketball player recently attacked a fan who heckled him, claiming he was negligent in the death of his daughter. The White man wanted the player to know who was superior. The athlete smashed the man's skull because he also wanted to show who was superior.

Rural people rarely have time to participate in sports. We have more important things to compete over, such as which farmer can grow the best stand of corn, which 4H child can raise the prize calf, or which logger can fell the most timber. Truckers compete by getting the best fuel economy, or raising the best children. City drivers who do not have to buy their own fuel often race to see who can get down the road the fastest.

About the only time most city folks come into contact with rural people is when they come out car driving. As in auto racing, they will often pass for no reason other than to express dominance or superiority. We are unwilling participants in a death sport. Sticks and stones may break my bones and names may never hurt me, but reckless car-driving might get me killed. Bigots often assault Blacks and homosexuals on sidewalks for the same reason — and are sometimes beaten themselves.

While rural people might be content with a rusty old pickup and be proud of their ability to keep it running, city folks consider a car's appearance to be the most important thing. They consider cars to be like lipstick, a hairdo, or an article of clothing — an indicator of their status in society. It does them no good to spend a lot of money on a pretty car if no one sees it, so they must drive recklessly and pass other people to be sure their fancy cars are noticed. Bucks snort and compare antlers just before they butt heads. If a city car-driver feels his superiority is being threatened he will often get stupid and and deliberately cause a collision. City folks who work in offices don't do meaningful, constructive, satisfying work which they can regard as a measure of personal accomplishment the way rural people do, so their status in society is based on appearance rather than ability.

A friend of mine just got in his first accident after nearly two million miles. He had stopped for a traffic jam, but left enough space that he could see the car in front of him. A city fellow driving a beautifully restored Fiat two-seater cut into his blind spot and when my friend started forward, they made contact. The car driver, who obviously was not competent to drive the car he owned, did not wait for the police and left the scene of the accident; then he called my friend's employer threatening to sue. He said the fender-bender had ruined the appearance of his car and he wanted the rural trucker to pay him money even though he was the one who caused the collision. The tragedy is that big city courts and juries often award damages for that.

I once hired a man who said he was a better driver than I was. "Oh yeah," I asked, "why then are you working for me rather than me working for you?" He told me how a car driver had committed suicide by cutting him off and slamming the brakes in front of his truck. "The stupid idiot knew I couldn't stop," he said, "he wanted me to hit him!" He told me how, after narrowly escaping the burning wreckage by jumping out of the door of his sleeper, he had been arrested and convicted of vehicular manslaughter. He lost his driver's license and was fired from his job with legal fees he couldn't pay — despite an almost perfect safety record. He said he had only one accident and two speeding tickets in five years!

The judge told him that because he was merely a truck driver, he should have sacrificed his own life by driving into a ditch to save a valuable member of the community. The conviction was overturned on appeal because the appellate court ruled it unreasonable to expect a person to sacrifice his own life to save another. That is a moral, not a legal decision. Besides, city folks aren't valuable: they eat good food; they don't do any honest work; and it costs money to flush their toilets.

The death was eventually ruled a suicide, but in the mean time the only job my employee could get without a license was to haul overweight scrap iron. He used back roads to get around the scales. Although he carried a gun to shoot the scale man if he ever got stopped, he chickened out and was fined for being 30,000 pounds overweight. (I stopped the truck immediately and told him to put his gun in the belly box.) He said they finally jailed him because there was no way he could afford the fine. In doing so, they did him a favor because once in jail his family was able to apply for welfare. The judge cut a deal whereby he would promise not to drive illegally in return for reducing his fine to $3,000 and get his license back. He was ordered not to see his wife and kids until then, except on the sly. That's how he ended up working for me. He was an honest man. He bought me a steak dinner because he felt I had tipped him too much and he would not accept charity. "Anything for another driver," I said, "you can pay me back." He said he had come too far to go into debt now. No one in his family had ever been in jail or on welfare before.

The constitution guarantees a trial by a jury. Most local governments are willing to provide a trial but not by a jury of peers. My peers are other professionals, not stupid idiots lacking the competence to back their cars out of their driveways. Most courts do not have the several hours necessary to educate jurors about the idiosyncrasies of trucks — even if truckers could afford expert testimony. Certainly a jury of truckers would never have convicted the man of manslaughter for running over a car that cut in front of him. Uninformed city folks hungry for revenge do!

I once had to take a load off a woman because she was rear–ended by a pickup truck, after stopping to check a tire she had just put on (as all good drivers do after 50 miles or so). The pickup slammed under the rear of her trailer at an estimated speed of 80-90 mph. This sheered the rear bumper clean off and shook the truck so much that it woke up her husband, who was asleep in the bunk. The guy in the pickup was decapitated. The cops arrested and cavity-searched her, then charged her with negligent homicide. They said she should not have stopped on the side of the road. Her more than 20 years experience had taught her to check new tires after a few miles because defective tires can fly apart and kill someone.

She begged them not to jail her because she was a grandmother and she could not afford bail. No one in her family had ever been arrested and she was worried about the example it would set for the grand-children. Naturally, they dropped all charges when the suicide note was found, but they never apologized. She was so shaken that she couldn't drive for more than a month and had to go on disability. It was not the collision that shook her up. She had seen many. It was the way the cops treated her. She said she shudders to think what might have happened to her if the family of the man had not come forward with the note.

Suicide is such a stigma that many people, urban and rural alike, keep such things secret in order to have fewer problems with life insurance companies. Because life insurance companies have often refused to pay death claims from self–inflicted injuries, car driving has become a leading form of suicide. When people survive slamming into a tree, telephone pole, or the back of a truck at 80 mph without their seat belt on, they always claim it was an accident. Most deaths connected with trucking are probably suicides, with or without notes.

Most car drivers are somewhat suicidal, though few admit it. Most people do not wear seat belts. I once worked with a man who was narcoleptic. He would fall asleep spontaneously even in the middle of conversation. After he totalled a company car, the Ford Motor Company checked his driving record and found that he had totalled two cars of his own that year besides the company car. Ford denied him the privilege of parking on company property. He filed a grievance with the union, saying that since the state of Michigan considered him competent to drive at 70 mph on public roads, Ford should allow him to drive at 5 mph in a parking lot. The plant's Industrial Relations Manager countered that since the car the man was driving was a Ford, the Ford Motor Company was a competent authority to make a determination as to the man's suitability to operate its products. The grievance was settled amicably when the company threatened to assign the man a parking space right next to that of the union steward.

When I asked the man why he continued to drive even though he knew he would be killed sooner or later, he said it was a risk he was willing to take. He said that life would not be worth living if he was stuck at home, unemployed with no form of transportation. The amount of suicidal nature in a person is related to the amount of risk they are willing to take.

His company will say he is a hero for saving a car full of kids. The media will say he fell asleep at the wheel. Other truckers will say he was just incompetent.

We Are Discriminated Against

We Are Discriminated Against

Why would the state of Michigan allow a person to drive at 70 mph. on public highways whom the Ford Motor Company considers unsafe even at 5 mph in a parking lot? Why should they deny a rural person the ability to earn a living just because they have epilepsy, diabetes, or have lost an eye in a farming accident? I once read about a Black man who won a two–year, quarter–million mile safe driving award — who had only had one leg. His local truck stop even assigned him a handicapped–only tractor-trailer parking space! The reason is that most people, including most politicians, are narcoleptic to some degree and few have only one eye or leg.

For years, because they are a minority, diabetics and epileptics were denied the right to earn a living even though their conditions were easily treated with medicine. Narcoleptics, because they are in the majority*, have been killing thousands of people a year for decades. They treat their condition by drinking large amounts of

---

* *55% of car drivers and 50% of truck drivers surveyed have had episodes of falling asleep behind the wheel.*

coffee, but ultimately their bodies get used to the caffeine and they must turn to harder drugs like ephedrine or amphetamines. When their system gets used to the drugs, the narcolepsy cannot be controlled. They will fall asleep and kill themselves or someone else if they are allowed to drive.

Even after killing someone, it is still legal for narcoleptics to drive. Millionaire politicians are reluctant to take away their own privileges — even when someone gets killed. Most truckers with more than ten years experience are, for obvious reasons, insomniacs. Narcoleptic drivers tend to get killed, fired, or scared out of the business within their first few years. Interestingly, truckers are victimized by hours of service regulations designed for narcoleptics.

Suppose a trucker, from a family of truckers with the genetic trait for insomnia, tried to obey the law. Lets say he starts work at 6:00 am and drives for ten hours:

>At 4:00 pm, the ten hours of driving he can complete legally are up and he is required to stop working for a mandatory eight hours of rest — so he watches TV at the truckstop until midnight;
>At midnight he is tired and ready to sleep, but his contract requires him to go back to work;
>At 10:00 am the next day his legal ten hours are exhausted again so he stops driving, but because it is morning, he can't get to sleep — he watches game shows;
>At 6:00 pm, he is required to go back to work.

At 4:00 am, he can finally get two hours of sleep before his 7:00 am delivery appointment — after 30 hours of driving and 46 hours of continuous on–duty time. No matter how tired he is, he must then unload his truck by hand! Is it any wonder that the most highly skilled professionals choose to violate the law?

If you are narcoleptic, you are are probably asking "why don't truckers just sleep in the daytime?" Insomnia is a genetic trait that runs in families — particularly trucking families. For myself, falling asleep behind the wheel is quite impossible. If I could fall asleep behind the wheel, I would be able to fall asleep in bed at night without first having a warm shower, a bowl of breakfast cereal, and a good book. I fall asleep easily only between the hours of 11:30 pm and 1:30 am. If I stay up much beyond Two, I usually head back out on the road. Driving makes me tired, so if I lose a night's sleep I sometimes can catch a couple of hours in the afternoon. I avoid doing that though, because it will throw my sleeping schedule off and I may not be able to sleep again until 3:00 or 4:00 am — not much sleep if I must deliver at 6:00 or 7:00 am.

One load required me to drive 24 hours continuously, stopping only for fuel. After 36 hours of continuous on–duty time without meals or so much as a moment's rest, a big dinner buffet and a warm shower in a motel room still couldn't get me to sleep. I watched TV until my usual sleep time of 11:30 even though I went more than 40 hours without sleep.

Going without sleep makes me angry and irritable. I daydream, not concentrating on what I am doing. An aggressive, angry driver can be even more dangerous than a sleepy one. Insomniacs need rest too. Getting enough rest and obeying the law are two different things. We are victims of genetic discrimination. In order to get adequate rest, I must go home or to a legitimate recreational facility. Since I started in 1988, my income has been cut in half and I can no longer afford motel rooms. I now work 100 hours per week, every week, from the time I leave the house until I get home. Turning down loads so as not to break the law increases the time I must spend between periods of rest. Sitting idle, watching TV away from home is simply not restful. Sitting still and doing absolutely nothing is hard work when you would rather be doing something else — ask any security guard or police officer. I am at my most dangerous when I have not been home for several weeks because I am angry about having to work overtime without receiving extra pay. For drivers with children, rest is impossible even if he rents a motel room; because he cannot know that his children are safe. For safety's sake, most drivers violate the law in order to drive as far as they can each day and get home as quickly as possible. For many drivers, obeying the law requires large amounts of money. How am I supposed to go off duty for a mandatory rest period if I am 2000 miles from home at some warehouse that doesn't even have a bathroom and I can't even afford a motel?

One time I thought I might fall asleep because I had the flu. No good driver would attempt to drive if he thought he might fall asleep, so I pulled into a rest area. There was a truck stop just ten miles farther down the road. I did not want to risk driving ten miles when I was tired. After I went to the bathroom, I was too weak to climb back into my tall Peterbilt 362 cabover. Another driver had to help me. I slept for two days, getting up only to vomit out the window. A cop and a prostitute tried to wake me; but I couldn't yell loud enough to get their attention. When my fever broke, I drove sixteen hours straight and delivered my load on time despite going two days without food.

Because of the gross disregard skilled professionals have for the laws made by millionaire–sponsored politicians, the government decided to subsidize truck–driving schools to replace us "safe" insomniac truckers with more patriotic, narcoleptic drivers who may or may not kill you when they fall asleep behind the wheel. If you have ever had breakfast at a truckstop, I'm sure you've seen drivers load up in the morning with multiple Thermos bottles of coffee, drive 70-80 mph. because they don't have to pay

WE ARE DISCRIMINATED AGAINST

Roll-over accidents are the leading cause of death among truckers.

Cargo must be laboriously transferred by hand to another truck before the fallen tralier can be righted.

for their fuel and do not care about fuel economy, then let you pass them up at every single rest area because their bladders leak from drinking too much coffee. Eventually they graduate to ephedrine, then amphetamines, then they fail their drug tests or kill people. I drive 55 (88kph.) and I'm usually the first in line to unload.

For many years, trucking companies refused to hire people who did not come from trucking families. A relative had to "get you in." This was the only way companies could be sure their quarter–million dollar vehicle would not go off the road into slumber–land. The government thought this was terribly unfair because Black people were discriminated against. Liability lawsuits kept drivers' wages high at a time when city folks were unemployed.

As the demographics of America changed in the post–war period and immigration swelled urban populations, political power moved to the cities. Shifting unemployment from urban areas into rural areas became politically feasible. Hitler solved the German unemployment problem with literacy driving tests so that Germans could take Jewish Jobs and shift unemployment into the Jewish community. Rural Americans have never had the good schools that suburban Americans take for granted. Literacy in rural areas is lower. American politicians were able to copy Hitler's method to shift unemployment into rural areas where it would not be as noticeable. Training urbanites to do rural jobs is no small feat. Taking jobs away from rural truckers who fail their literacy test would just raise freight rates if urbanites cannot be trained to replace them. The problem is that urbanites are not truly American. The original Americans emigrated to escape religious persecution and were willing to endure any hardship in order to live their lives as they chose. City folks immigrated to the United States to get rich, to live a life of leisure, and do as little work as possible. Anyone can be trained to drive a truck; but few are willing to work one hundred hours a week the way most rural

66

people do for their entire lives.

Whatever the type of training, it had to be quick. City folks like instant gratification. They don't understand the concept of Spring planting for harvest in Autumn. A couple of city folks who bought my videos wrote that they wanted their money back because the videos did not show them how to shift gears — as though they would start killing people by actually driving an eighteen–wheeler ten minutes after watching a video about it.

Trucks, unlike cars, have manual gearboxes of a type that used to be known as "crash–boxes" because they do not have synchronizers to automatically engage the gears. The gears must be clutched manually or "double–clutched." The technique is to take the truck out of gear at the exact moment that you let up on the accelerator, shifting into neutral, then "feel for the hole," letting the rpm's drop while applying pressure on the next hole until the vibrations stop and pushing it into the next gear without depressing the clutch pedal at all. Novices, because they have not yet developed the feel, must use the clutch pedal twice with every shift — first to take it out of gear and then again to put it in at the right time, hence the term "double–clutching." In a car, you push in the clutch only once and the synchronizers engage the next gear automatically as soon as you push the gear lever into the next hole.

Once in a while a friend will ask if he can drive my truck. "I dunno; can you?" I always ask? "I've never driven one and I would like to know what it is like," he says. "Well, that kind of proves you can't," I chide. The reason trucks have crash transmissions is that they are lighter, more durable, last longer, do not wear out your left leg, and most importantly, most people can't drive them. Trucking companies cannot tell from a ride around the block whether a person is a good driver. It is unlikely that a novice would be put into an accident situation that would test his ability on so short a drive. Trucking companies, therefore, test whether new hires can shift gears. If you can't shift smoother than an automatic; get an automatic! You have no business driving a truck.

Enter the truck driving school — which promises to do in two weeks what most drivers take two years or longer to accomplish — becoming qualified to drive an eighteen–wheeler. Truck driving schools could best be described as gear–shifting schools. The object here is to fool the trucking company into hiring the novice by impressing them with gear–shifting ability. I broke down once and asked a driver training truck for a lift. The instructor sat quietly in the right seat as the student ground every gear. We were driving on a freeway service road devoid of traffic. The sum total of the driver training was to speed up, slow down, and go through the gears every quarter mile.

Safe driving has nothing to do with operating the vehicle. My mother has driven her car four hundred thousand miles without a chargeable accident. My father, who taught her to drive, had a half million miles. He was taught to drive by my grandfather — a professional driver. The reason I am a safe driver is because my parents were. Of the eleven expert drivers selected for the American Trucking Association's 1994 Road Team, five were sons of truckers. The reason most people are not safe drivers is because they too were trained by their parents — who drive worse than they do.

When I was hired at North American Van Lines for my first trucking job ever, the manager of our department said to all of the government subsidized truck driving school graduates, "we are going to do drug testing, so if you are using any illegal substances, you may as well leave now." A third of the graduates got up and walked out. Then he said "we are going to examine your driving record in all 50 states, so if you have had any accidents or tickets you may as well leave now." A man stood up and said, "I've had an accident; most people have, but I've also got a $3,000 dollar student loan to repay. How am I supposed to repay that if I can't get a

Trucks are the only vehicles manufactured that are not required by federal law to have roll-over protection. They are also the only vehicles <u>required</u> by federal law to be top–heavy.

A truck tailgates a car. Do not judge him too harshly. He is driving exactly the same way most people drive their cars. That's the problem!

A student truck driver hooked his trailer tires on a curb while leaving the fuel island. Instead of looking in his rear–view mirror to find out why the truck

would not move, he just gave it more gas. The tires exploded with the force of two sticks of dynamite, causing everyone in the truckstop to run out and see if their trucks were all right. His lead foot cost his instructor nearly $1,000.

69

job?"

"You've already proven that you can't drive a car safely; why would you expect us to hire you to drive a truck?" the manager answered.

Of the 120 at the interview, only 14 of us were invited to attend the training program. Of the three that were eventually hired, none had been to truck driving school, and all of us had related experience. I was a security courier and fireman at Ford. One was a co–driver and the other had straight truck experience. How many truck driving schools have you heard of that flunk 75% of their students? Trucking companies are very selective.

Politicians creating diploma–mills will not create additional employment in the industry. The amount of freight to be hauled is dependent on the amount of products people are willing to buy. People do not eat more food just because truckers are unemployed. Can you imagine parents asking their children to eat twice as much broccoli to help unemployed truckers that haul broccoli?

Why Politicians would train drug users to drive trucks, I cannot understand. The Federal Student Loan Program was created to send Black people to college — not to send drug users to truck driving school. I would like to suggest the politicians knew all along that the established companies would not hire the graduates. My theory is that about the time Ronald Reagan took power, the Republicans had a problem. The problem was that every time a Black man graduated from college, the Democratic party got another campaign supporter. Truckers, with their $60,000 per year incomes, were strong Democratic supporters already. Training people to do work that they would never be hired to perform would bankrupt the Federal Student Loan Program and prevent Black people from going to college, and the few driving school graduates who were hired would lower drivers' pay so that experienced drivers could no longer afford to financially sup-

70

port Democratic candidates — killing two birds with one stone, so to speak.

The reason truckers hate graduates is that if their daddies have money, they can take a shortcut to get into the good jobs, whereas most of us earned the right to sit in the driver's seat by working our way up from the bottom. The reason most truckers today are White is because 20 years ago, most teamsters were White. Trucking companies hired the children or relatives of teamsters to serve an apprenticeship, starting with lumping or unloading freight. If they showed the right attitude toward safety, driving a fork lift truck safely in the warehouse where their boss could supervise them and spotting trailers in the parking lot, they would be invited to drive straight trucks where their boss could not supervise them. Only after a couple of years parking trailers and driving straight trucks would a new driver be considered qualified to drive an eighteen–wheeler.

Today, 90% of lumpers are Black, therefore we can assume that by the year 2000, 90% of truckers should be Black; but that is not going to happen. The truck driving schools have created a glass ceiling for these up–and–coming Blacks. Why would a trucking company take the time to train a Black Man when they could hire a White man who had been trained for free?

I once hired a man who was an Imperial Dragon of the Ku Klux Klan. He was a tireless hard worker, so I gave him a tip. He said it was too much and invited me over to his home for supper. When I asked him why he had revolvers all over the house, he told me it was to defend himself from the Black Panthers who were out to get him.

"But you got along fine with that Black man I hired," I exclaimed; "does he know you're in the KKK?"

"Of course not," he answered; "he's a good nigger; I've hired him many times myself."

"How can a KKK man hire Blacks?" I asked.

"I'm an affirmative action employer!" he exclaimed.

When I asked how an Imperial Dragon of the Ku Klux Klan could be an affirmative action employer, he explained that the Klan was only against Whites working for Blacks, not against hiring them. "Blacks *should* be made to work for the master race," he said, correcting my assumptions. He continued that any businessman who did not hire the best–qualified applicant was selling himself short and would go out of business in the long run. His wife, also a member of the Klan, nodded her head in agreement.

If the Klan is willing to hire qualified Blacks over unqualified Whites, then there must be some even more virulent, self-serving, hard–core racists in the government who want even worse than the Klan to prevent Blacks from raising themselves up the ladder to success.

Enter the Job Training Partnership Act. If legitimate, established trucking companies will not hire unqualified applicants, then the government will simply subsidize new companies that will run the established companies out of business. To describe how this works, let us pretend you are an unemployed businessman who has never done an honest day's work in your life, and your daddy is a millionaire who makes big contributions to Senator So–and–so's election campaign.

"Why don't you start a trucking company?" the Senator suggests at a campaign fund-raiser.

You explain to the senator that you don't know anything about trucks and you do not have any money to start a company with.

"No problem," the Senator retorts, "you can capitalize the company with taxpayer's money!"

"Aren't big trucks dangerous," you ask?

"Not for the person who sits behind the desk," chides the

Senator.

So, on the Senator's advice you start a dummy corporation that exists only on paper and lease some trucks from an equipment leasing firm. Equipment leasing firms are familiar to little old ladies whose husbands leave them with a lot of money that needs to be invested in a safe, secure income stock. The financial planner or broker will take four or five percent of the widow's money as a front end "load" to compensate him for setting up the "limited partnership." As partners, little old ladies are judgment–proof. No jury can be convinced to take their retirement savings away, even when they profit from the killing of innocent people. With a rented building, some office furniture, and a leased computer system, you can be in the trucking business without any attachable assets. This is important because when you hire people to drive forty–ton trucks who aren't even competent to drive the cars they own, you are going to kill a lot of people and you are going to be sued. You must be ready to liquidate and set up business in a new state at a moment's notice if the jury awards against you become unmanageable.

Leasing rather than buying things is expensive. Fly–by–night trucking companies would not be able to compete against established companies if it were not for government subsidy. The Job Training Partnership Act pays these "kiddie–car" companies to train their drivers with taxpayer's money. The justification for these subsidies is that the students would not be able to pay back their student loans if the Government did not throw good money after bad in order to get them hired.

Millionaires who invest in trucking can make big money by paying their drivers less than minimum wage. There is a special exemption in the Fair Labor Standards Act for transportation workers paid by the mile. With a flood of graduates looking for work chasing a limited number of fly–by–night jobs, a student can be made to work 100 hours per week and only drive a thousand miles at 20¢ a mile — only $200 dollars per week! A company which pays $13 per hour plus another $12 in pension and health care benefits cannot compete with a company that pays less than minimum wage with no benefits and is subsidized by the government. When the expert drivers laid off by the established company try to apply for work at the kiddie–car company, they are turned down because there is no training subsidy for hiring experienced drivers. The world's best drivers are replaced by incompetent, drug–using ones and huge profits are made — which can then be kicked back into political campaigns!

I once met a driver who worked for Shaffer Trucking (the same company as myself) who said he was not making enough money. Shaffer, an established company in business for forty years, had been bought out by a millionaire. The driver claimed to have worked a hundred hours a week for only $132.00. "Let me see your log book," I demanded. He showed me that he logged twenty hours a day in a 36–inch–wide sit–in sleeper! "Not even Dracula could spend so much time in a coffin so small," I shouted; "if I were a DOT man (Department of Transportation inspector), I'd give you a ticket for falsifying your logs — this is unbelievable!" He told me someone in our company's management had told him to log all his on–duty–not–driving time as sleeper–berth time so that they would not have to pay him for all the time he spent on duty. I told him to log all his time as on–duty–not–driving whether he was working or not and march into our boss' office and demand to be paid the minimum wage. I understand he was fired or quit shortly afterward. There would be a lot of employee–owned trucking companies if they were required to pay their drivers' back wages.

Suppose you went to a town where there were two hospitals on the same street and you asked under your breath, "Why does this town need two hospitals on the same street?" So you walked into the first one, with old, ivy–covered walls, and found very old, experienced doctors inside — overworked, tired, and haggard look-ing, operating with obsolete, broken down equipment in dirty hallways. So you decide to go to the other hospital. The other hospital is brand new and clean, with shiny brand new equip-ment, but there are no doctors in sight. You approach a school–kid standing behind the counter and ask, "I'd like to see a doctor."

"Like, I am a doctor," the kid answers.

"Aren't you a little young to be a doctor," you ask, half thinking he might be some kind of child prodigy.

"Like, I just graduated from doctor training school two weeks ago."

"Well, how much training did you have?"

"Like, it took twelve whole weeks, man."

"Do you have a license to practice medicine?"

"Like, the millionaire who owns this hospital like, had to give Senator So–and–so a big campaign donation to get me my license."

"What kind of operations can you do?"

"Like, I can do anything, man: brain surgery, heart transplant, kidney dialysis — I can do it all...They gave us a com-puterized virtual reality expert system that we put on our head that like thwonks us when we screw up."

"Why did you decide to become a doctor?"

"Like, I was planning to go to truck driving school, but like they had doctor training school across the hall; like doctor training school graduates get to make minimum wage — not truck driving school graduates."

As you get back in your car to go back to the first hospital, two trucks come up behind you. One truck is dirty, old and worn out, leaking oil, but without a scratch on it and there is a tired hag-gard looking driver behind the wheel. The other truck is shiny and brand new, just out of the factory, but with a big dent in the bumper and the radiator grille missing. Just then the new truck tailgates and rear ends you. As the ambulance crew straps you into the stretcher, a young kid gets out of the truck and says, "don't worry mister, a friend of mine just graduated from government subsidized doctor training school the same time I did...He'll fix you up!"

Ideally, the finest professionals should have the finest equipment. Public hospitals should not be old and run down and the trucks that million–mile–without–an–accident expert drivers own should not be leaking oil. Just as the best doctors should have the best hospitals, the best drivers should have the best trucks. If minimum wage, kiddie–car companies bid freight rates so low that drivers cannot afford braces for their children's teeth, some expense is going to have to be cut — not the children's braces, but maintenance of the truck. Instead of buying a new truck every five years as was traditional, trucks are now purchased only when they are totaled in crashes. Old trucks break down more often; but experienced drivers can do many of the same repairs as mechanics, so minor repairs tend to be neglected until the driver arrives home.

In order to prevent expert drivers from remaining in business, the government began a system of federally–sponsored state truck inspection programs. Rather than simply alerting the driver to defects so that he could fix them, state inspectors impound the vehicle so that any perishable freight is completely destroyed if the driver does not hire a government approved mechanic to repair the defect on the spot at several times the cost of repairing it himself.

I once met a driver who had been coerced into buying a $300 steer tire for $600. They charged $480 for the tire and $120 to mount it. Normally, mounting a tire would cost $20. When I looked at the old tire, it had nothing wrong with it except some cosmetic cupping, which was normal for the type of truck he was driving. The tire had several times the DOT minimum tread depth of 4/ 32 inches. He said the officer had measured the depth only in the cupped portion.

An officer in Iowa once wrote me a warning because my steer axle brakes were down to just a quarter–inch of lining. While it is true that rear axle brakes have rivets a quarter–inch high, steer axle brakes have rivets only an eighth–inch high. The linings are only three–eighths inch thick even when they are are brand new. According to the officer's manual, the brakes should be replaced even with half their design life left in them. By this standard, car and pickup–truck brakes are illegal even when they are brand new!

When a driver wrote to Overdrive Magazine that wrecker drivers in Oregon and Washington were gouging truckers at a rate of $600–900 per tow, a trucker who drove a wrecker there wrote in a few months later that he would tow anywhere in the state for $350 and that only the government–approved wreckers were price gouging.

There has never been any proof of widespread kickbacks to Department of Transportation inspectors; but the price gouging is hard to explain any other way. One time in the Washington State Port of Entry on I-5, a DOT man asked me if he could have one of my shirts. Shaffer Trucking has a cute little bird mascot printed on it's uniforms named, "Percy the Puffin." Several of the other inspectors had hats they said had been given to them by other drivers. The officer followed me out to my truck and I got a brand new shirt out of my suitcase.

I said, "That will be $4.25."

The officer's jaw fell open.

"It's a brand new shirt that has never been worn and that's what I paid for it," I told him. He was obviously upset as having to pay for the shirt and gave me only nickels, dimes, and quarters.

In Massachusetts, a law was passed requiring brake adjustments to be performed by qualified mechanics. Brake adjustments are a highly personal thing to truckers. It is sort of like packing your own parachute — your life depends on it so you want to make sure it is done right. Even if a mechanic with 30 years experience adjusted my brakes, I would still have to get underneath and do it again myself to make sure it was done right. Brake mechanics are trained in the same truck driving schools that the kiddie–car drivers are. Why should I pay the State of Massachusetts $120 to hire someone to do what I do myself practically every week and then have to do it myself again less than ten minutes after they do it in order to make sure that it is done right?

While it is true that the brakes of a fully loaded truck should be adjusted as tight as possible, the trailer brakes of a partially loaded truck should be backed off so that they will be in balance with those of the tractor. When I worked for Mayflower, I used to pull a tandem–axle trailer with a single–axle tractor. The weight on the drive axle was twice that of the weight on each of the trailer axles. If the trailer's brakes were not adjusted to have half the braking effort of the tractor's brakes, the trailer's wheels would lock up before the tractor's brakes were even on halfway.

This can happen even to a tandem–axle tractor if the partial load is stacked near the ceiling because trucks are top–heavy and the weight tends to shift forward during braking. The problem got so bad at Mayflower that they required us to buy our own trailer tires to install on their trailers. The problem wasn't just that $800 worth of tire rubber was destroyed in skids; sliding tires provide less braking effort than rolling tires do and adjusting the trailer brakes as tight as possible as required by law causes a type of accident called a "jackknife" in which the tractor starts moving slower than the trailer and the trailer swings around in front, wiping out five or ten cars. These accidents are not caused by a collision with any other vehicle. They are caused by ordinary, commonplace application of the brakes when the trailer brakes are adjusted legally. Government regulations deliberately cause accidents.

You might think politicians are crazy for requiring expert professionals to risk their lives needlessly. Actually they are not, because no expert professional would ever attempt to obey this law. Trucks are taller than they are wide — thirteen and a half feet tall and only eight and a half feet wide, making them very top–heavy. Should the truck slide into a ditch in the jackknifed position, the truck can roll over, with the entire weight of the tractor and half of the weight of the trailer resting on the soft fragile cab. Politicians know we would rather pay a fine than to risk our lives. Truck inspections are designed to raise revenue, not to increase public safety.

Every type of vehicle in America is required by law to have roll–over protection, cars, farm tractors, fork lifts, and bulldozers — except for tractor trailer trucks. It is just not feasible to design a cab that can support 30 tons. Straight trucks and delivery vans do not usually have a problem with roll–over. Any cargo compartment which can support 20 tons of cargo can support 10 tons of up–side down truck as well. Dump trucks usually have an extension of the dump body over the cab to provide protection. When cargo is stored exclusively in a trailer, there is no protection for the cab and the driver can be crushed. All a driver can do to save his life in a jack-knife is release the brakes and accelerate — even if it means deliberately bashing into cars.

I first got the idea to produce videos about trucking at the scene of a roll–over accident. A truck rolled because of a pavement defect. There was a one to two inch drop–off at the edge of the con-crete on the cloverleaf exit ramp to northbound 322 from eastbound I-81 in Harrisburg, Pennsylvania. Pennsylvania is con-sistently voted in the Overdrive Magazine Trucker's Survey as having the nation's worst roads. The lane was so narrow that in order to keep our left trailer tires from tracking off the pavement, it was necessary for the driver to position his right steer tire precariously close to the drop–off. As the lane veered to the left, the driver's right front tire fell off the pavement and was jerked to the right by the rut. The lane was banked into the turn on this 45mph. high speed cloverleaf. The asphalt shoulder, negligently, was not level with the road and dropped off from two inches beneath the pavement at the edge to two feet beneath just a couple yards away. As the truck's right front tire jerked to the right, the other right side tires of the truck followed and went down the bank, dropping about a foot lower than the tires on the left side, which were still on the pavement. As the right tires suddenly became a foot and a half lower than the left tires, the roof of the trailer (because trucks are twice as tall as the wheel–base is wide) flung abruptly three feet to the right in less than a second. This flinging together with the centrifugal force of the truck going around the turn at 40mph. caused the truck to roll. The driver said he tried to get back under control but he could not climb back up the two inch drop off until after the truck was already past the point of no return. He kept on steering even as his truck was sliding on its side.

I nearly hit the car in front of me because it slammed on its brakes when the truck rolled. I immediately ran to help the driver, who could walk even though his head was covered with blood. The ambulance crew strapped him in very securely as the lacerations made him look as though he might pass out at any time.

When the fire crew arrived they started digging up dirt to soak up fuel that was leaking. I pointed out to them that, according to the manifest, the truck was loaded with kitty litter which would be much more effective at soaking up the spilled fuel. With halberds, they chopped a hole in the trailer roof and passed in a pretty, petite female firefighter who began tossing out enough bags of the stuff for us to soak up all of the spilled fuel. The safety director of Howard Transportation, (of Laurel, Mississippi) who owned the truck, said he would buy me a steak dinner for doing that because toxic spills are sometimes the most expensive aspect of accidents these days.

When television reporters showed up, the attitude of the police changed. While the plainclothes sergeant thanked me for helping out, his uniformed subordinates suddenly wanted to do a truck inspection at the exact time that I was showing the reporters the drop-off at the edge of the pavement. They were not only afraid the reporters would blame the state of Pennsylvania for the accident, but one of the officers (together with some firefighters) had stolen a naked lady air–freshener and called the driver a pervert, and I threatened to tell the reporters that too.

The officers made charges against the driver they knew were untrue. Based on the man's logbook, they deduced he had driven a thousand miles that night — all the way from Kentucky. He had told me he had only driven 20 minutes prior to the accident and that he had spent the night at the Carlisle Truckstop. The brakes were stone cold, proving what he told me was true, but the cops charged him with excess hours anyway. I got really upset when they charged him with illegal possession of a firearm. The Second Amendment of the Bill of Rights in the United States Constitution grants an unconditional right to bear arms. Truckers who live in their trucks have the same right to defend themselves when they sleep as any homeowner.

Guns have been a necessary part of trucking for three hundred years. One safety official at Howard remarked over the phone that he believed two–thirds of his drivers carried guns and that any driver with a quarter–million dollar truck and cargo would be a fool if he didn't. One driver told me if it were up to himself he would not carry a gun, but his wife insisted and bought him a pretty chrome Derringer for his birthday. Drivers cannot just call the police when they become the victims of crimes. The cops might give us a $1,000 logbook fine for daring to bother them. Drivers have no choice but to deal with criminals on their own — vigilante style. That is why bodies are occasionally found in truckstop lots. Truckers almost never tell on each other; just helping at an accident scene might get you a $85 fine — that's what happened to me!

First they told me to stand in one place and not to talk to any reporters. I did not recognize that they had any legitimate authority over me because I was on foot, off duty, and my truck was locked. I was no different than any other pedestrian. When I went over to talk to the reporters, the officer who stole the air freshener got out and threatened me with disorderly conduct — just for talking to reporters! I gave them 20 days worth of logs and they spent twenty minutes looking for errors without finding any. They charged me with "logs not current" because I jumped out of the truck without changing my duty status. They said I should have completed my federally–required paperwork before rendering aid to the injured driver.

Having spent two years as a security guard/fireman at the Ford Motor Company, I know that ambulance drivers do not sit and complete their paperwork while people are injured — though as professional drivers, the law would seem to require them. The officer said I should have entered the name of the town I ate breakfast at twice instead of just once and I should have entered the time more accurately. I told him that it was none of the government's business where and when I eat breakfast and to record the time I left any more accurately, I would have had to remove both hands from the steering wheel and both feet from the floor, and turn my head completely sideways to write at 60 mph!

Of course, I was guilty. Not a single trucker has ever complied with the law. No one would try to write in a logbook while driving a 40–ton vehicle at 60mph. Since every person subject to the law is guilty of violating the law, the law is discriminatory. It violates our first amendment rights because the only way it is enforced is selectively. A thousand equally–guilty truckers went by on I-81 while I helped the firefighters soak up the spilled fuel. The only reason I was fined $85 dollars was that I stopped to talk to reporters.

I was no more or less guilty than any other trucker on the road that day. Worse than that, if I had been politically correct and told the reporters that Pennsylvania has fine roads and that all police officers are heroes, I would not have been fined. I was fined because I told the reporters there was a dangerous highway condition in their town that could kill someone at any time and that a cop had helped steal a five dollar pornographic air freshener. Logbook laws are unconstitutional, not because the public does not have the right to know when a trucker stops to go to the bathroom, but because the selective enforcement of these laws can be used to violate our first amendment rights of free speech.

What is surprising is that the two officers involved were Black. The plain-clothes sergeant who thanked me for helping was White. It seems that being a victim of discrimination, as with child abuse victims, makes a person more likely, not less likely to discriminate against others — just as victims of child abuse are more likely to abuse their own children. Discrimination is an on–going social cycle where previously oppressed groups rise to power only to oppress other groups. The Puritans, who were persecuted by the Anglican Church while in England, in turn persecuted the Quakers here in America. The problem is not with Blacks or the KKK; it is with a top–heavy, overbearing government that seeks to micro–manage the day–to–day lives of ordinary people and give strangers power over us.

After watching the reporters work, I thought that I too could carry a video camera. I bought a home movie camera and mounted it in the back of my sleeper so I could film without stopping. Two years later I produced "So You Want to Drive a Truck?," the first and most successful film about trucking ever to be produced by a driver. (The film can be ordered by sending $19.95 to Trucking Video, Box 4265, Sargent Texas 77414.)

Before I ever drove a truck, when I lived in a Black neighborhood, a Black man started chasing me. He was faster than me and was gaining on me even though I had on a brand new pair of expensive running shoes. Since it was inevitable that he would catch me, I thought that if I dropped my wallet he would just take the money and leave me alone. As the wallet hit the ground he shouted, "Mister, hey mister, you dropped your wallet!"

I stopped immediately and waited as he brought me my wallet, then asked, "Why are you chasing me; I thought you were a mugger or something!"

He laughed so hard he could hardly catch his breath and exclaimed, "Me a mugger, ha, ha, ha, I thought you were a cop!"

Then I started to laugh. "Why were you chasing me?" I repeated.

"I'm on the high school track team and I'm trying for an athletic scholarship; I was just looking for someone to run with."

"Why can't you run by yourself," I asked?

Then he got very serious and said, *"You know what the cops do when they see a Black Man running."*

Discriminatory laws affect people in subtle ways that changes our behavior, often without us knowing or thinking about it. Truckers used to be called the "Knights of the Road," helping disabled motorists in need. The next time you are hurt or broken down at the side of the road and a thousand trucks go by without even one stopping to help, remember the logbook fines the drivers risk if they dare to stop.

The Knights of the Road have been replaced by a "new breed" of trucker. Rather than rural family men, these city folks are likely

84

to be on drugs or perverted — people who can't get legitimate jobs anywhere else. As expert professionals are driven into bankruptcy, so are the mom–and–pop truckstops that depend on their business. The giant "shopping mall" style truckstops that replaced them come complete with male and female prostitutes and drug dealers. You can't take a shower in a common shower room for free anymore. The shower rooms have to be cleaned after every shower. The attendants must now be paid to remove condoms, enema syringes, and human feces from the floor. You don't dare take a shower without a pair of beach togs on.

Once, I was having one of those all–you–can–eat steak dinners you get after you buy an oil change when I noticed a little girl crying in the next booth over. She was perfect, with big blue eyes and long blond hair down to her waist covering a light blue little girl top with her belly–button showing and white short-shorts. She could have been a fashion model. Her pimp was throwing her a birthday party and the truckstop had provided the cake. Judging by the candles, she was fourteen. It was her first birthday away from home. There is no telling how many men used her long smooth legs that very night.

It takes millions to buy elections. Politicians had a problem that we truckers were beating the brains out of student drivers for cutting us off, weaving out of their lanes, driving too slow, and otherwise driving their 40–ton trucks as recklessly as they used to drive their cars. After they killed hundreds of people, the millionaires thought we were too violent for trying to stop them and decided to get rid of us by cutting freight rates and driving us into poverty. They said we were stupid for buying $60,000 homes back when our incomes were $60,000 a year. We should have guessed that congress would vote to deliberately cut our pay by half while simultaneously doubling their own pay. They needed no cattle cars to take us into the country the way Hitler did to the Jews. We already live there in poverty, out of sight, except for little children who wonder when their daddies will come home.

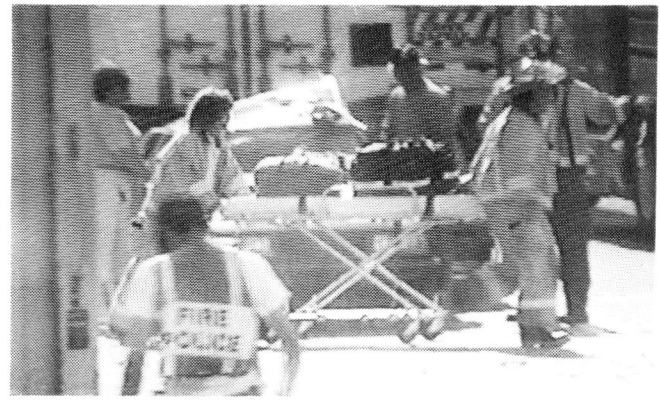

Paramedics rescue crash victims.

Traffic Laws Are Stupid

Traffic Laws Are Stupid

Most people drive cars the way they play chess. One player says, "Check." The other says, "No problem, I'll just make a sudden evasive manuver...I have lightning–quick reflexes!" The first player says, "Checkmate." The other says, "That's not fair...I didn't see it coming!"

The art of defensive driving can be compared to playing chess. Rooks and pawns don't cause much trouble; their behavior is predictable. The bishops and queens really must be watched out for. They zig-zag all over the place. The knights are the worst of all. They jump in front of you through places you'd swear they'd never fit — then slam on their brakes.

The highway is like a chessboard with your own vehicle being on a black square and with red danger squares in front, beside, and behind. Other safe black squares are kitty–corner to yours. The principle of safe driving is simple: stay out of the other guy's dangerous red squares. As long as vehicles remain 100 feet apart, collisions cannot occur. Before a collision can occur, vehicles must first come within ten feet of each other. Before they can come within ten feet, they must first come within 100 feet of each other. As long as vehicles remain 100 feet apart it is impossible for them to collide.

The reason so many collisions occur today is that the father of the modern interstate super–highway system, Mr. Adolph Hitler felt that inferior people like you and I should stay out of the way of pure Aryan aristocrats like himself. After all, his three–axle Hispano–Suiza limousine could do 100 miles per hour and the Gypsy donkey carts could only do five or six at most. When he built the first autobahn, (German for freeway) he included passing lanes so that superior races could pass inferior slower people. To this day, there are no speed limits on the Autobahn and Volkswagens must by law get out of the way of BMWs. The inferior races do not always get out of the way in time, so collisions on German freeways are quite spectacular, involving hundreds of cars.

In America, we believe that all men are created equal with certain inalienable rights, such as the right to life, liberty, and pursuit of happiness. Statistical studies have shown that if everybody traveled at the same speed there would be almost no collisions. If everybody stayed 100 feet apart and traveled at the same speed, collisions would be impossible.

Unfortunately, there are too many descendants of Europeans in America who feel they must prove their genetic superiority by passing everybody else. If they run over a child chasing a ball into a street, they say it is the child's own fault for playing in the street. I cannot remember how many times stupid people have honked or shouted that I should ride my bicycle on the sidewalk, knocking over little old ladies, when I was going just as fast as their cars were. Would you like to see a cyclist zooming inches away from small children at thirty miles an hour?

92

TRAFFIC LAWS ARE STUPID

If the author can ride his ten–speed bike across the United States, you can certainly ride one to work.

In suburban neighborhoods, children are told to stay out of the street so as not to bother the neighbors. Children should be seen, not heard. In high crime neighborhoods (usually Black or Hispanic), parents tell their children, "Be sure to play in the middle of the street where the neighbors can keep an eye on you." I was once an amateur bike racer and I rode my bike across the entire United States. Because driving a car safely is completely beyond most people's ability, they believe we minorities (including cyclists) should all join monasteries and never come out in order to avoid being killed!

I once encountered a European who apparently did not realize that light signals in the United States mean the exact opposite that they do in Germany. He flashed his high beams in my mirror, blinding me, indicating that I should change lanes in front of him — which I obligingly did to avoid being blinded again. He slammed on his brakes and attempted to pass on the right; but then gave another signal that I should move right. When I did so, he again slammed on his brakes and attempted to pass on the left. He then signaled that I again should move left. Then he again changed into the right lane and signaled that I should move right. Finally, he passed me normally without any more signals, accelerating as fast as he could. I realized about halfway through the escapade that he must have come from Europe where shining high beams into the rear view mirrors of others means "get out of the way!" Shining bright lights in people's mirrors is not only rude, it is dangerous. If the man signaled me to stay in front of him I was happy to do so; because as long as my trailer was between my mirrors and his lights, he could not blind me.

Once a stupid idiot told me that because the windshield of my truck was five feet higher off the ground than his car windshield was, that it was not necessary to dim his high beams when my truck approached. I became so mad that I grabbed him by the neck and cracked his skull against a wall with a resounding thud. I always wondered what I would do if I ever caught one of these people that cause ten–car pile–ups. He was lucky I did not hurt him worse.

The theory behind speeding laws is that everyone should travel at the same speed and that lower speeds result in less severe collisions. The problem with speeding laws is that while traffic may be safer at a lower speed, an individual car is safer at a higher speed. The faster you go, the more control you have over your surroundings. By traveling faster than the traffic you can position your king at the safest spot on the chessboard. Because all the other stupid idiots try to do the same thing, the safest spot is always changing. The speeders jockey for position to try to keep from being hit.

Statistically, speeders do not get in any more collisions than those of us who drive at a constant speed. Speeders tend to slam into slower–moving people. Every time a speeder is involved in a ten–car pile–up, many non–speeders are as well. Speeders do not like to slow down because they feel vulnerable to being hit by other speeders. Speeding gives them a feeling of control. They speed because they are frightened.

Speeders say they are above average drivers. They endanger the rest of us by weaving through traffic like some out–of–control life–and–death video-game, but because an innocent person is hit every time a speeder crashes, speeders are in no greater danger of being injured by speeding than those who are hit by them. The chances of hitting someone and the chances of being hit are about the same. "Why shouldn't we speed," they argue, "we don't have to pay higher insurance rates than non-speeders do — reckless driving is fun... Driving is an enjoyable sport... Sports-cars are sporting equipment... Demolition Derby is a sport!"

It is not an ethical sport because most people are forced to play involuntarily. You wouldn't enjoy boxing if someone forced you to do it against your will. You would have a hard time convincing car-crash victims suffering in the hospital or survivors at a crash victim's funeral that cars are sporting equipment. Car companies do not care if their cars are safe or easy to drive. They only care that they are thrilling and exciting to drive in the five minutes it takes to sell one. Most customers are timid, cautious, and a little nervous when riding around on a test drive with a high–pressure salesman. If the controls seem over-responsive, they think that the car is peppier and more thrilling than the one they own.

Ideally, cars should have no more than 40 horsepower and at least four inches of accelerator pedal travel so that power can be added at a rate of no more than ten horsepower–per–inch of pedal travel. To make the cars exciting on the test drive, car companies equip them with two–to–three hundred horsepower and only two inches of pedal travel so power is added at a rate of 100 horse-power–per–inch of travel. Is it any wonder that most car drivers are not competent to drive the cars they own? They are constantly speeding up and slowing down, unable to control their speed to match the speed of other traffic because their accelerator pedals are overly sensitive. The car might accelerate or slow down abruptly just because the car goes over a bump or if the jolt changes the driver's foot position.

In order to obey speeding laws, car drivers must watch their speedometers constantly. They spend half their time looking at the dash and less than half watching where they are going. They rarely ever have time to look in their mirrors. To travel at a constant speed, they must use a dangerous and inappropriate device called cruise-control which causes them to slam into things without slowing down, providing power to the wheels even after a collision. To disengage the device, the incompetent motorist must lift his or her foot off the floor — something quite impossible with the car spinning wildly out of control!

Steering wheels are also overly sensitive, with only two to three turns lock–to–lock with a 90–degree wheel cut and a 15–inch steering wheel. Trucks have four and a half turns lock–to–lock with a 60–degree wheel cut and a 22–inch steering wheel. If the car driver so much as leans while taking his eyes off the road to look at his speedometer, his car can drift completely out of its lane. It almost seems as if car companies deliberately design cars to have collisions just so they can sell more cars.

I say, the best thing you can do with your speedometer is smash it with a hammer. There is a saying in the trucking industry: "Show me a driver who is watching how fast he is going and I will show you a driver who is *not* watching *where* he is going." That is why radar detectors are so important as safety equipment in trucks. Trucks need three times the stopping distance cars do. It is imperative that truckers keep their eyes on the road at all times. If you need to know whether or not you are going too fast, you should look out the window. If you are approaching within 100 feet of another vehicle and that vehicle is traveling at a safe speed, then you are going too fast. If other vehicles are passing you recklessly then you are going too slow. Since few truckers know or care precisely how fast we are going from moment to moment, we need radar detectors to avoid constantly getting tickets when we drive safely.

If you have ever looked inside a race car, you may have been surprised to find almost no gauges. Professional driving demands absolute concentration. If there are any gauges at all, there will be a tachometer and an oil pressure gauge sitting on top of the dash where the driver will not have to avert his eyes to see them. Trucks need other important gauges such as a pyrometer, water temperature, and air, manifold, and brake application pressure gauges.

These will be in the top row because they can warn of a cat-astrophic breakdown that could cause a collision. Less important gauges such as amps, voltage, speedometer, gearbox and axle temperature gauges should be underneath.

Speeding tickets constitute an unfair and probably unconstitu-tional tax on safety since speed enforcement forces professionals to divert their attention to a less important part of the instrument panel at a time when car drivers are driving recklessly. Drivers are fined for simply attempting to drive safely — keeping their eyes on the road when looking down at the instrument panel would be bad judgment. Split speed limits, when the speed limit for cars is dif-ferent than for trucks, require truckers to drive at dangerously slow speeds. Even the safer car drivers are forced to risk their lives and pass our 70–foot long vehicles or become an obstacle to others approaching from behind. Drive dangerously long enough and far enough and you will get killed eventually. Trucks are 70 feet long. It takes several seconds to pass one. If an obstacle comes up in the road ahead, the truck will need to change lanes whether a car is in the way or not. The most common cause of fatalities connected with trucking are lane change accidents.

I once read in a newspaper about a bizarre incident where an unwed teen–age mother was trying to get rid of a baby she did not want. She was riding in the back of a pickup and threw the baby under the wheels of a truck. Truckers will not run over any object — not even an empty shopping bag because it could be full of cement and cause a ten–car pile–up. Many states have $1,000 lit-tering fines for good reason. Volunteers who pick up litter are heroes because they save lives. The trucker swerved and the baby, who was probably knocked unconscious in the fall, passed under-neath, between the tires. A car-driver, who was tailgating in an effort to get slightly better fuel economy did not make any effort to avoid hitting the object and ran over the infant. Infants are thick,

TRAFFIC LAWS ARE STUPID

A little girl rides unprotected in the back of this pickup. The driver was driving on the shoulder so I could pass.

A car aggressively crosses the double white line to cut off the author's truck in Chicago. Such a bold move will make no difference in the impact speed. If his engine even sputters, his death is certain.

A car cuts dangerously close in front of the author's truck in Houston.

so the car was launched into the air and the car driver lost control, ending up in the ditch.

The trucker called the police on (CB) channel 9 and chased the pickup until a cop could intercept. He then made a citizen's arrest for reckless endangerment, thinking that the mother was deliberately trying to wreck his truck by throwing an unknown object under the wheels. The car driver who hit the infant, upon removing the obstacle from the road to prevent others from also hitting it, discovered, much to his horror, that he had run over a human being. It was the police dispatcher who put the two stories together and ordered the patrol officer to charge the mother with murder.

A friend of mine passing through downtown Reno Nevada, once heard a call on the radio, "Someone just ran over Bambi!" Another voice answered, "Does Bambi wear a red shirt and blue-jeans?" Then he saw blood spattered all over the road and swerved into the passing lane to get around it. Cars passed him on the right, running over the body and spattering his truck with blood. Once decapitated and squashed flat, the drunken man's body did not provide much of an obstacle. My friend said he witnessed more than sixty cars, nine out of ten that encountered it, run over the corpse without giving it a thought.

If nine out of ten people will run over a body without thinking, that means that nine out of ten people who work where you do, or nine out of ten people who sit next to you in church are also likely to run someone over without thinking. What do these people say to God when they drive their cars to church — that they have to go to church so badly they do not care if they kill someone? Do they say, "I don't care if I have to kill an entire family, I'm going to church — no matter what!"

Before I became a driver I was an opinion researcher, and Friday night was a particularly good night to do opinion surveys.

People had a lot of free time on weekends. I had just left a Catholic neighborhood where there was a Virgin Mary in front of every house, and was just entering a neighborhood strangely without any statues when, as if a silent gong had rung, hundreds of people came out of the houses wearing funny black suits with unusual short trousers that let their socks show. The ladies had scarves and long dresses. The men had pancakes on their heads. They passed me as though I wasn't there — entranced like ghosts. While I was a little annoyed at their unwillingness to be polluted by contact with goyim, I realized that these Orthodox Jews were better than Christians. They, at least, were unwilling to kill on the Sabbath. Their religion prohibited them from driving cars or operating dangerous machinery. If only they could be so considerate about driving during the rest of the week. A riot was caused in New York when a Hasidim ran over a Black child. The charges against him were dismissed!

There was a man in California who had a marital dispute with his wife. He swore that if he did not get custody of their son, no one would. When the court awarded custody to the woman, the man took his son to a motel room, gave him a sleeping pill, covered him with kerosene, and lit him on fire. Fortunately, the motel had a fine smoke–detection system. Firefighters were summoned immediately who arrived in time to save the child's life; though he was burned over 90% of his body. The man was caught and sentenced to life in prison.

In New Jersey, at about the same time, a man wanted a pack of cigarettes. He loved his child so little that, not wanting to hire a baby–sitter, he took the child in the car with him even though he knew he was not a competent driver. As the car caught fire after a predictable wreck, he ran cowardly away, not even thinking to rescue his son.

One boy was burned because of an insanely angry marital dis-

CREATION OF A MINORITY GROUP

pute; the other coldly for a pack of cigarettes. One father was charged with attempted murder and got life in prison; the other was not charged with any crime because authorities considered it an accident that the little boy had fallen into the car with him. Both boys were burned over 90% of their bodies and scarred for life; why shouldn't the punishment be the same for both fathers?

I entered a construction zone on I-80, near Davenport Iowa once, doing about 25 mph, when I noticed a car speeding along the median, passing all the other cars backed up in traffic. As I entered the cones, the car suddenly accelerated, then slammed on its brakes as the stupid idiot inside realized that there was only thirty feet between my truck and the car in front of me — not enough room to slow down at 80 mph. Rather than simply driving through the cones, which are made of rubber and would not have harmed his car at all, he chose to drive underneath my trailer, a maneuver with a 100% chance of certain death. As if to prove to all who were watching that driving the car he owned was completely beyond his ability, he locked up all four wheels — doubling his car's stopping distance. His wife, who I could see in the passenger seat through my mirror, put her hands in front of her face, silently screaming as his skidding tires screeched. By some miracle, he stopped just short of hitting the reefer fuel tank under the trailer, missing both trailer and cones by inches. I drove onto the shoulder to avoid hitting his car with the trailer wheels.

As the backed up traffic began to move, he darted behind my truck and started passing me on the shoulder, driving onto the grass when a reflectorized pole got in his way. As I went over a narrow bridge, he could go no farther and again darted behind me. When traffic again stopped, he got out of his car and started shouting something, shaking his fist. He appeared to be foolishly challenging me to fight even though I was armed.

We eventually came to a cop, whom I alerted by flashing my

102

A car tailgates to get better fuel economy in Dallas.

Moments later, a car aggressively crosses the double white line, cutting off the truck behind me.

Did they enjoy their vacation? The children in the back of the pickup were thrown out and killed when it rolled, smashing the Fiberglas top near Barstow, California.

103

headlights on and off rapidly. The cop got out of the car and I pulled off onto the shoulder, holding my hand to the windshield, pointing to the left where the cop could see it. As soon as the way ahead was clear, the man chirped his tires, accelerating as fast as his car would go. The cop quickly took the wind out of his sails by motioning for him to pull in front of me.

When I got out of my truck I overheard the cop say, "I've heard about as much of your lip as I'm gonna take...one more word and I'm gonna give you some time to calm down!" Nodding to the woman, he asked, "What exactly is the problem here?"

"I want to press charges," she said.

"Against your husband," the cop inquired?

"No; against that truck driver," she answered, pointing an accusing finger at me.

"What did he do?"

"He's trying to kill us," she shouted, weeping!

The cop asked me what I did and I shrugged my shoulders and told him about the reckless driving. A couple of construction workers came forward to collaborate my story. The cop asked us if we wanted to press charges; but said he would have to let them go if we did not want to return to Iowa to appear in court. None of us could do that so he told them they could leave. Trying to get the guy arrested, I stuck my head in the wife's window and asked, "How can you drive that way with children in the car with you?" He predictably started cussing and swearing, but the cop was Black and did not arrest him. I doubt that a White police officer would have put up with such disrespectful treatment.

The cop said that there was no point in arresting the man because he had a nice car. "People with nice cars have nice lawyers that get them off," he said; "it's not worth the paperwork." Officers are frustrated by the limitations imposed on them by the courts. There is a double standard of justice — one for the rich and one for the poor; so much so that a pure Aryan White can be disrespectful to a Black even when he is a cop. With good lawyers, the rich can do whatever they want. The cop will have to take out his frustrations on somebody else — such as truckers.

There is, on average, about one death for every million miles of truck driving. The average trucker goes three million miles in his career. I believe the Second Amendment about freedom of religion gives professional drivers the right to drive safely even when the law requires otherwise. The Bible says "Thou shalt not kill." We have a human right not to kill.

The only speeding ticket I have ever received in more than a half–million miles of trucking was given to me in Oregon. It was foggy, with patches of ice and I was descending a hill in reduced gear. As I emerged from a cloud–bank my radar detector went off. The brakes had been on since the top of the hill and I saw no reason to worry. Two other trucks that had passed me moments earlier were between myself and the cop, and I assumed he was shooting at them. As I went up the next hill, I lost sight of the cop as I went into another cloud bank. When I arrived near the top of the next hill, flashing lights appeared out of the fog signaling me to pull over. I was doing 35.

"Was I going too fast or too slow," I asked as I climbed out expecting he would give me a 'too fast for conditions' warning.

"Too fast, you were doing 78!"

"You're out of your mind, that's ten miles an hour faster than my truck can go!"

"I clocked you on radar."

"Lets see it."

"I don't have to; a lot of people think we have to show you the (radar) display but in Oregon we don't."

"How much experience do you have using that thing?"

"Twenty–five years; I've been using radar since there was

radar."

"You realize my company will fire me for going that fast; they have a strict policy about leasing trucks that go more than 70."

"That's not my problem," he said, as I got the distinct impression he was trying to extract a bribe.

"I'll see you in court."

"Okay, I'll write the court's phone number on the back of the ticket for you." I drove five miles to the town the court was located in and called the court clerk.

"I need to see the judge immediately, a crooked cop just wrote me a speeding ticket!"

"I'm sorry but she (the justice of the peace) will be busy doing arraignments all day and she's got a wedding at 2:00 pm."

"I'm entitled to due process; I have a right to a trial; weddings only take five minutes; I thought JP's were supposed to be available during normal business hours!"

"Not in Oregon...I can schedule a trial date."

"A trial date to see a justice of the peace? It would cost me two thousand dollars in lost income and travel expenses to do that!"

"Well then, why don't you just pay the fine?"

"I'm innocent! That wouldn't be just. How much is the fine?"

"$270."

"I can hire an attorney for less than that."

"Why don't you do that then."

After calling more than ten attorneys, I found one younger than I was who would handle my case.

"Who's on rotation as prosecutor this week?" he asked the Court Clerk over the phone.

Making another call: "Can you be DA (District Attorney) to handle a speeding ticket?" He told the other attorney the specifics of the case. "Will you plead guilty to 68," they asked?

"That's faster than my truck can go! I'm not going to admit to

Trucks have very little visibility on their blind right side. If you think the cars in these mirrors look small with the camera two feet away, imagine how small your car looks to the driver seated eight feet away.

Most truck driving students say they want to learn to drive trucks in order to see the country. These scenes are pretty now, but wait until winter!

107

Unloading thousands of boxes stacked floor to ceiling by hand from a modern tractor–trailer is beyond human endurance. Drivers are often forced to drive after being forced to do this work — which makes them too tired to be safe.

This driver hired two lumpers for twice the normal rate. When they took too long to unload, he gave them less than the agreed amount, so in retaliation they pulled his kingpin release as he was getting into his truck. The trailer fell onto the ground, causing serious damage. If the trailer had been loaded, it would have broken in half and cracked the concrete. This warehouse required lumpers to put the freight into a cooler 500 feet from the dock — significantly more than the fifteen feet allowed by law. The lumpers felt justified in charging more because of the extra work involved and because they

had worked as fast as they could.

committing a crime that is impossible. I haven't gone faster than 64 since I entered the state of Oregon."

"He says he'll plead guilty to 64."

"That's too big a drop; its 68 or no deal," he says; "Can you go 66?"

"I dunno, I might have gone 66."

"Good, just take these papers to the courthouse and pay them $270 in cash. You will get a refund of $170 in six months. That'll be $210, you can pay me by check."

"$210, you've been working for me for less than an hour!"

"I saved you $170."

"I have to pay $210 to save $170?"

"I picked you up at the truckstop."

"I'm a professional and that's more than I get paid for driving."

"I saved your job."

"I was innocent; you made me plead guilty."

"Okay, $170, that's the way the system works."

"Now I know how a Black man feels. I'll pay you a hundred bucks an hour plus 25¢ a mile; that's $100.75."

"$175, you're wasting my time."

"I thought the initial consultation was free."

"Not if we do something for you."

"That prosecutor was a buddy of yours, $125."

"$150."

"Shake; Perry Mason you are not, good buddy," I said to him as I wrote out the check.

Though I wasn't speeding before I got the ticket, I had no choice but to do so afterward. The whole experience put me six hours behind schedule. It was the day before Thanksgiving and the drop yard would be closed if I didn't make it there by midnight. I had to take the transmission out of gear to make better time on the hills. Going faster than seventy with the engine in gear would over-rev the engine, damaging the rockers and possibly throwing a rod. Out of gear, speed is unlimited.

When I got to Denver, my brakes were hot from rolling down hills with the truck out of gear and a bunch of idiots had decided to park their cars in the middle of the interstate! Cars should be required to exit the highway immediately whenever their speed falls below 45mph, but most people feel their own personal Thanksgiving trips are more important than other people's safety. The cars in back could see the smoke coming from the trailer wheels but the cars in front could not. When they cut me off and slammed their brakes on, I went for the shoulder.

When they saw me passing them on the right, they must have said to themselves: "Hey, that truck is taking cuts in our line; well, two can play at that game; we can take cuts too," so a group of them figuratively put their guns to their heads, playing Russian roulette with a big truck without brakes — cutting me off on the shoulder with nowhere to go when a steep embankment on my right and cars parked on my left prevented me from maneuvering. Fortunately, I arrived at the bottom of the hill before I ran over any of them and I was able to get into granny gear to keep going until the brakes got enough cool wind to be able to use them again. The fifty gallons of extra fuel I burned was for nothing because I missed my Thanksgiving dinner anyway. Thanksgiving dinner that year was a hamburger, and not a very good one at that.

Most of the time when I find myself driving recklessly it is the fault of the customer. Lumpers who unload trucks have a 300 year history in the trucking industry; but this obsolete system of patronage perpetuates because millionaires who own warehouses like to save money by denying employees health care and retirement benefits. They pay their employees through an indirect money laundering scheme through trucking companies and truck drivers

Fork lifts require very precise driving — positioning cargo within an inch of where it is supposed to go. This is the best way to learn to drive a truck because in the warehouse, drivers can be supervised and the potential for damage is less.

Modern fork lifts have "space shuttle" controls that make the transition to trucks more difficult. One of the best methods of training drivers is now being lost.

A man holds his daughter in a truck.

111

that can best be compared to washing money in the drug trade. Should anything go wrong in the negotiations necessary to hire a lumper, unloading might be delayed and the driver put behind schedule and therefore be forced to drive recklessly. When you consider the parcel service, the pop truck, or the vending machine guys, with their 20–50 deliveries a day, it would seem silly to expect their customers to come out and help them unload their trucks. Small businessmen cannot afford to hire a guy to stand around all day doing nothing, waiting for a single package to arrive. Naturally, the driver is expected to unload it and bring it to the door.

If a driver pulls two 53 foot long turnpike doubles, unloading 90 thousand pounds of freight by hand in one afternoon is beyond human ability. I can unload 400 one–hundred–pound boxes of meat (40,000 pounds) by hand in about three hours. Two thousand ten–pound boxes of frozen pizza (20,000 pounds) takes six hours even though the load weighs only half as much. A load of more than 3,000 boxes cannot be unloaded by a single driver in a single workday no matter how light the load or how physically fit he is.

I had the misfortune of hauling a load of 4,000 ten–pound boxes of toothpaste into Lucky Stores in Fullerton California. Lucky does not allow Blacks. Though we were in the middle of a Black neighborhood, there was only one Black face in the building, a supervisor in a different department. When I told them I couldn't possibly unload the truck myself, a woman gave me an approved list of lumpers who were all white or Hispanic. When I called and told them what I had to unload, none of them wanted the job. A white guy agreed to work for me for $25 per hour — triple what a Black man normally charges. When the man arrived I quickly found out why no one else wanted the job. The supervisor gave my lumper four bills of lading for four different Lucky trucks bound for four different destinations around Los Angeles. My lumper was told to sort the load between the trucks. He did not physically have to load the trucks; just stack the load on pallets by hand so the pallets could be loaded onto the trucks in just a few minutes by fork lift.

Many think that because drivers hire and pay the lumpers that lumpers work for the drivers. Actually, it is the other way around. Lumpers are warehouse employees. The driver may pay the lumper, but the supervisor or checker tells the lumper what to do and the lumper tells the driver what to do. My lumper at Lucky worked three part–time jobs. He said he had to leave at noon.

"What am I supposed to when you leave," I asked? I could not interpret the Lucky documents. I told him to stop sorting the load and simply unload the truck since there was no time to do anything else. He said he had to do what he was told or he would be fired, or his name removed from the list of approved lumpers. I told him that for $25 per hour he should do what I said. "Where else," I asked, "could he get a job that paid so well?" He explained about costs to work there and that he didn't get to keep all the money. The floor supervisor got a split. It took me nine hours to finish the job he started.

When the truck was empty, this supervisor refused to sign that he had received the load. He said I had to re-stack some of it since I had not sorted it properly to be loaded onto their trucks. I told him that was warehouse work, nothing to do with unloading the truck and it was illegal under ICC regulations to coerce a driver. He said "Here's a quarter; go call the police." Suddenly realizing why there were no black lumpers, I was flabbergasted! Most grocery warehouses have a whole gang of lumpers standing outside the gate waving and smiling and yelling "I'll unload that truck for you!" But not Lucky. They must have paid the police to keep them away. One would think that in south central L.A. there would be whole armies of young Black men willing to work for $25 per hour tax–free cash money. They should be knocking down the gates to get in. How else could Lucky prevent a driver from hiring a Black man off the sidewalk if the police were not involved?

I met a Black lumper at a different warehouse sometime later who said he had worked at Lucky. He said he would never go back because they wanted him to re-stack the load three times and do twice the work that the white and Hispanic lumpers were required to do. This incident took place shortly before the riots where several Lucky stores were looted or burned. You could say the Black community taught Lucky a lesson; but I have not been back to see if they learned their lesson. Any Black who shops at Lucky is a traitor to his or her race.

Lucky put me five hours behind schedule, so I missed my load. A team (truck with two drivers) had to be diverted from San Francisco to Tulare to cover it. That meant I had to deadhead 410 miles that night to cover their load. After doing backbreaking work for more than nine hours, I had to drive a further eight hours for a total of 17 hours of continuous on–duty time without stopping for breakfast, lunch, or dinner, or so much as a coffee break. My hands and wrists were so tired from gripping the little toothpaste boxes that I couldn't hold the steering wheel. I had to put my hands and wrists through the spokes and steer with my elbows. I had to drive 750 miles on each of the next three days to deliver the team's load on time. I had no time to stop for meals or get an oil change that was overdue. I bought snacks to munch while driving.

Millionaire politicians, wealthy beyond our dreams, seem to have the absurd notion that trucking companies can get away with crapping on their customers. Why, if I was so tired, didn't I just pick up the load a day late, you may ask? The reason is that in any business, a contract is a contract and promises must be kept. A business that routinely breaks its contracts will not remain in business very long. My company has contracts with its customers

It is illegal to take my mother on a trip with me, but I do not get to see her very often, so I break the law.

The owner of this trailer is obviously very religious, but the driver will probably get a ticket if he tries to go to church.

When police started harassing truckers for parking in residential neighborhoods near churches, missionaries responded by bringing

churches to truckers. Even if the truckstop owners do not give permission to use the tv room, these mobile chapels can park in the lot on Sunday morning to give truckers a chance to worship.

115

to pick up loads on certain dates. If we break those contracts we will lose those customers.

I remember when our company pissed off Hershey Chocolate and another company, KLLM, won our contracts away from us. My income fell in half. Instead of profitable long hauls of 1,000 miles or more I was doing short hauls of 500 miles or less and the deadhead (driving with the trailer empty to go get the load) was sometimes longer than the haul! There is no amount of speeding or logbook fines the government can levy that would persuade a company to deliberately screw its customers.

The government subsidies of truck driving schools and the training programs of "kiddie–car" trucking companies did not just replace good drivers with bad ones. By forcing good companies like mine to compete with the bad ones, all drivers, good and bad, including myself, are forced to put safety concerns aside as the price of remaining in business.

When I was in high school, my teachers threatened to go on strike. They said it was not fair that truck drivers should make twice as much money as they did. Truck drivers, they said were high school drop-outs. Teachers had Master's degrees. To me, the argument seemed silly. School teachers do not risk their lives every time they step into a classroom. It takes just as long to teach a driver to drive well as it does a teacher to teach well. If we can shorten a truck driver's apprenticeship to two weeks, then maybe we should shorten a four–year college to four weeks, law school to eight weeks, and medical school to twelve weeks. Would you want a doctor to operate on you who was only twelve weeks out of high school? Then why would you want a truck driver driving behind your car who who had only two weeks of training? The government subsidizes training programs to keep trucker's incomes low. Low incomes mean low freight rates and low freight rates mean millionaires get richer at the same time as you and me get killed.

With so many drivers illegally taking their families with them, many are trying to clean up the airwaves so their children will not have to listen to city folks "talking trash."

This driver has let his tires wear much too far before having them capped.

Because police listen in on CB conversations, drivers must invent "handles" or code names to protect their anonymity. Trucking companies must make lists of their drivers' handles if they want to communicate with them.

117

Low income sometimes means we must do more work than we should. I had a load of pizza to deliver to Albertson's in North Salt Lake City, Utah, where the lumpers wanted $20 per hour. I offered them ten. They refused, so I did the work myself. All I had to do was break down the tops of the pallets because the boxes had been stacked too high. The lumpers threw wads of paper at me and broken boards into the back of my trailer as I was trying to unload it. The worst they did was block my aisles in the warehouse where I was supposed to put my freight. Sometimes not all of a product is found in the same place in the trailer. We have to put the partial pallet aside and then put it back in the trailer when more of the same product is found. My partial pallets were completely blocked by cargo from other trucks.

When my truck was empty the checker refused to sign that he had received the pizza because he said it had not been sorted properly. He wanted me to re-stack it. I told him that sorting the product is warehouse work having nothing to do with unloading the truck and coercing a driver to do extra work he is not being paid to do is a violation of the law. He said, "So what! You're not getting out of here until the load is stacked so we can receive it!" The argument was moot since neither he nor I could get at the product with all the freight blocking it. After two hours of waiting for them to move the freight I went to the plant manager and waited another hour for him to come out of his meeting. Finally, a supervisor was dispatched and the checker signed me out immediately. The checker told the supervisor I had been harassing the lumpers.

The three-hour delay made me miss my load and a team had to be dispatched to cover it. I had to cover the team load 2,500 miles from Twin Falls Idaho to Brockton Massachusetts in three days. I did 800 miles for each of the first two days, but in the third day I still had 900 miles to go and you cannot make good time in the New England states no matter how good a driver you are. I had to call my dispatcher and tell him I was going to be late.

"Dammit, that's the third one this morning I have to reschedule. I guess I'll have to fuck the drivers because I don't have time to talk to you. All I've been doing all day is rescheduling appointments."

When I got to the consignee they asked me why I was late. They had been told my truck broke down. They said I should get a new tractor if it is breaking down all the time. They refused to unload me until after all the other trucks were through as a punishment for being late.

You are probably wondering why trucking companies put up with customers like Albertson's and Lucky. The reason is that P&G and Pillsbury are willing to pay the lumper fees no matter how high they are. They just charge more for the toothpaste and pizza next time. Lucky saves money by not having to pay its employees' health care and retirement benefits. The lumpers make money because Blacks would have their jobs if it were not for the police. The police make money because protection rackets are a long–standing tradition. It is illegal as hell. Truckers have the right to hire whom they please. Blacks have the right to stand on a public sidewalk and ask for work. Truckers have a right to have their delivery receipts signed after a load is delivered without being extorted to do warehouse work for which we are not paid.

The reason we all don't go to the police or the ICC is not just that the police are corrupt; there is a blame–the–victim mentality built into the system the politicians are trying to perpetuate. Any driver who complains about being forced to work excessive hours would get a fine for working excessive hours, just as a driver who complained about being forced to haul an overweight load would be fined for hauling an overweight load. The multimillionaires who own Albertson's and Lucky donate huge amounts of money to

congressional campaigns. The fines exist to keep the victims silent.
Suppose the governor of a state announced that his state had
solved the problem of rape — and that there had not been a single
rape reported since his state had solved the problem of women
being raped. When a reporter asks, "how did you do it?" The governor says:

> "It was simple; all we do is put a large metal object
> called a speculum inside women's bodies when they claim
> to have been raped. Most virgins say this is worse than the
> rape itself. If the woman was not a virgin, we make her tell
> the names of all the lovers she has had or we put her in
> prison for contempt of court. Ever since we started jailing
> women and setting the rapists free not a single woman
> has claimed to have been raped."

This is a true story. This is the way rape victims used to be
treated before woman's rights marches, boycotts, and bra burnings
made the politicians change their minds. People become politicians
because they crave power; they want to rule over others. Rapists
rape because they crave power; they also want to rule over someone
else. Politicians and rapists are snakes of the same stripe.
Silencing victims with arbitrary fines and punishments
perpetuates corruption.

A police officer tailgates another motorist.

Hypocrisy

Hypocrisy

I f you drive a truck long enough and far enough, you're going to see bodies. It's inevitable. Most people aren't competent to drive the cars they own. Most people will kill themselves if they drive their cars long enough and far enough. Fortunately, most never drive far enough in their lifetime for death to be a certainty. Your life depends on beating the odds.

Most collisions are caused by sober, unimpaired drivers. Roughly one in ten Americans is an alcoholic, just as about one in ten Americans is Black. Statisticians expect that since one in ten cars is driven by an alcoholic, about 20% of two–vehicle collisions and nine out of ten ten–car collisions will involve an alcoholic even if his or her intoxication is not pronounced enough to have an effect on driving ability. These statistics also hold true for Blacks. If one in ten cars is driven by a Black person, two in ten two–car collisions and nine out of ten ten–car pile–ups will include a Black person. Since drunks are involved in 40% of all collisions, 20% can be attributed to the effects of alcohol and 20% would have happened anyway even if alcohol intoxication had no effect on driving ability. To say that 40% of all accidents are caused by drunks just because 40% involve drunks would be like saying that 90% of ten–car pile–ups are the fault of Blacks just because 90% are thought to involve Blacks. Not all collisions that drunks get into are caused by the drunks themselves.

If a speeder hits a well–lit drunk driver advertising his presence by weaving from side to side, how many other things might the speeder hit which are not well lit, such as deer, fallen trees, or children on bicycles? Eighty percent of all collisions are caused by people who drive their cars more recklessly than drunks even though they are sober.

Many people think that they can understand reckless driving by analyzing car accidents. If that were true, one would expect that police officers would be safe drivers. Officers crash into things, on the average, once every 28 thousand miles — three times the accident rate of truckers even when the infamous government–subsidized truck driving school graduates are included in the statistics. Aggressively passing at more than 100 mph. during a high–speed chase demonstrates a blatant disregard for human life. The usual reason given for such chases is that drug dealers and car thieves must be caught at all costs, as though a wealthy person's stolen car is important enough to risk ordinary people's innocent lives. The real reason officers chase is that they do not want the criminals to be able to defy their personal authority. Their self esteem is based on the assumption that police officers are superior to social inferiors who must be made to obey. Their personalities are so aggressive that admitting defeat and asking others for help in apprehending the criminals later, when they can be arrested safely, makes them feel inadequate. In this regard police are no different than any other speeding motorist that is late for work.

I do not mean to imply that police officers are below average drivers. Compared to car drivers, police are above average. But, this book is about professional drivers. As professionals, most officers are not competent. If a trucker were to be caught by his safety director driving at 100 mph like a typical police officer, he would likely be fired on the spot, or at least required to attend remedial training before being allowed to drive again.

Some non–trucking viewers of my first video, "So You Want to Drive a Truck," wrote that they could not understand what the police officers did that was so reckless because they themselves drove that way every day. A survey was taken in which car drivers were asked whether they considered themselves above average or below average drivers. Ninety percent claimed they were above average — a statistical impossibility indicating that 40% stupidly believe they are above average when in fact they are not. Most people overestimate their driving ability. A rookie cop, a few weeks out of police academy, killed himself when he slid his patrol car sideways into a Houston skyscraper. His car hit so hard it folded in the middle around a post and bent 90 degrees, like a boomerang, with the front of the car pointing in a different direction than the rear of the car was going. Police investigators estimate he had to have been doing over 100 mph., even though the speed limit in downtown Houston is only 35 mph. There is no indication as to what he was chasing or why he was going that fast. It is amazing that he would attempt to make a turn at that speed.

So, why is it that so many police officers drive unsafely? Maybe it's their attitude. Police officers have to have an aggressive attitude in order to enforce the law. You wouldn't want a police officer to just stand by and do nothing when a crime is occurring. You want that police officer to rescue you from the criminals. The problem is that a person who has an aggressive personality usually isn't a safe driver. Aggressive people tend to drive aggressively, get mad, accelerate, and slam into things. Even though police officers investigate more car crashes than anyone else, they are not necessarily safe drivers.

Significantly more cigarette smokers electrocute themselves in the bathtub than non–smokers. This is not because people smoke while taking a bath. Only careless people drop their electric shavers and hair dryers into the tub while bathing. With all the publicity surrounding the risk of cancer and heart disease from smoking, it can fairly be said that only careless people tend to smoke these days. Careless people smoke as well as electrocute themselves. There is nothing in cigarette smoke that causes electrocutions. It is carelessness, not smoking that causes electrocutions. Careless people electrocute themselves regardless of whether they smoke.

Careless people tend to drink more than most people. Excessive drinking is caused by carelessness. Carelessness causes car accidents. If a statistician adjusts for when the accidents occur, taking into account the fact that most drunks drive at night and the chances of hitting something in poor visibility is greater at night than in daylight, the odds of crashing while drunk are only slightly higher than driving at the same time of night while sober. Most drunk drivers do not get caught drinking and driving because they simply slow down when their reaction time increases. If getting drunk doubles the reaction time of a sober person — who then drives at half speed to compensate for slower reflexes; the odds of hitting something going slowly while drunk on a road devoid of traffic are no greater than driving at normal speed while sober. "Careful drunks," who drive slowly, will not cause any more collisions than sober people. Careless drunks, on the other hand, drive just as fast while intoxicated as they do when they are sober (some even faster because alcohol makes them more aggressive when intoxicated). These are the most careless motorists of all; but

A Texas trooper passes on the right with "disco lights."

Trucks get damaged in car crashes, too. It is unlikely the car driver will pay for the damage.

This truck, owned by a big trucking company, has its hub cap held on with duct tape. Oil can be seen leaking from the hub. Should it all leak out, the bearings will seize and the wheel will come off, causing an instant jackknife and possibly a pile–up. A self–employed trucker would see his truck

impounded for this kind of condition, but big companies seem to get away with anything. This same company once threatened to sue me for showing in a video one of their trucks making a delivery with two flat tires on one wheel.

129

unless they are extremely intoxicated, it is the carelessness, not the alcohol that causes the accidents.

The assumption is that if alcoholics would just stop drinking, highways would be safer. This idea is deceptive because careless people will still be careless even when they are sober.

By harassing drunks who drive too slow, police unwittingly increase the number of collisions by forcing drunks to drive faster than they want to go. Most states have a "too fast for conditions" clause in their reckless driving law which requires motorists to drive 10 mph slower than the posted speed at night or during poor visibility. A drunk doing 35 mph at night on a road posted for 45 mph, is actually traveling at the maximum safe speed. A person doing 45 at night on a road posted for 45 is actually speeding by 10 mph. Nevertheless, it is the drunk, and not the speeder who is harassed by cops.

Many people have reaction times slower than drunks do, such as handicapped people and Senior Citizens. Teen–agers may have quick reflexes, but locking up all the wheels is not the reflex one wants if one is trying to get stopped in an emergency. While it may be acceptable to require teen–agers to ride bicycles or take a bus to protect the public from their poor driving ability, most Seniors are unable to do so.

I know a couple of stroke victims who take more than five minutes to get out of their car. A bus full of passengers cannot wait that long just for one person to climb aboard. Many people have to drive. They have no other choice.

A law that requires people with slow reaction times to pass a test of their reflexes prior to obtaining a driver's license would prevent most Seniors from driving even before most Alcoholics would be affected by it. Like Alcoholics, Senior Citizens have good days and bad days. Test a person when their arthritis or rheumatism is acting up and they may have an extremely slow reaction time. Test them after they take their medication and they may test near normal — just as a Drunk would if he had time to sober up before the test. Many medications taken by Seniors are stronger than alcohol and actually cause drowsiness. A law that banned driving under the influence of drowsiness inducing drugs as well as alcohol would affect the Seniors more than the Drunks.

Because politicians allow absolutely anyone to drive, regardless of whether or not they are competent, it is not safe for Senior Citizens to ride a bike, a horse and buggy, or merely to cross a street corner without a Boy Scout. If all people with slow reaction times are banned from driving, we will rob Seniors entirely of their ability to travel outside their homes and function as active participants in society. Seniors have important contributions to make, most notably to their grand-children. Would you suggest that we risk the lives of children to go see their Grandma? It is better to risk Grandma's life than to risk the lives of little children.

It is imperative that we make the highways safe for all motorists, regardless how slow their reaction times may be. If we make highways safe for Seniors, they will also be safe for Drunks. Speed limits must be set low enough that even Senior Citizens can drive safely. Speeders who subject others to risky passing maneuvers just to save a few minutes on their trip must be charged with reckless driving; and if the reckless passing is not justified by mitigating circumstances: be convicted of a felony and lose their licenses.

Studies have shown that Drunks and reckless drivers often continue to drive even after losing their licenses. As with Senior Citizens, there is frequently no alternative transportation for them available. Bicycles are the safest form of transportation in the world. They are also the most dangerous form of transportation in the United States. Bicycling is dangerous in this country because politicians let absolutely anyone drive cars and trucks whether they are qualified or not. Then they selectively discriminate against some bad drivers who happen to drink while allowing other equally bad drivers to go on driving, risking the lives of law abiding bad drivers who choose to use bicycles as an alternative to continuing to drive without a license. The obvious unfairness leaves little moral incentive to comply with the law. No matter how bad a motorist may be to deserve to lose his license, he is still safer driving his own car illegally than to risk his life on a bicycle sharing the road with millions of other equally bad drivers who have not yet been caught.

In Communist China, only professionals are allowed to drive. To drive an ordinary six–wheel straight truck, one must attend a four month training course and endure a year–long apprenticeship. Hitting a bicyclist is tantamount to assault with a deadly weapon. Drivers who hit things are severely punished. It is hypocritical to expect a drunk or other bad driver to significantly increase the danger to his own life by riding a bicycle just to lower the danger to other motorists who drive worse than they do. In the United States, driving a car while drunk out of one's mind is still safer than bicycling.

It is discrimination to revoke the licenses of drunk drivers without doing the same to all the others who drive equally badly. Many people, especially those from Europe have traditions of alcohol use in the home. In France and Italy, wine is served at every meal. The English drink ale. The Germans drink beer. Hillbillies and Westerners used whiskey to purify water. If a Drunk or Senior Citizen breaks the law by driving recklessly, the loss of their license should be part of the punishment for breaking the law. Robbing people of their ability to actively participate in society just because their blood alcohol level is too high, when they have violated no other law, is cultural bigotry.

A few drunks are good drivers. One of the instructors who

taught me to drive had two million miles without an accident. He also downed a six–pack of beer over lunch. Many drunks in rural areas have never had an accident. They drive slowly; and there are few obstacles on country roads to collide with. Alcoholism is a disease that requires compassionate treatment — not punishment. Isolating people from society just because they happen to drink just prevents them from coming forward for help.

One time when I was homeless, I saw some activity at a church so I went in, thinking I might crash a wedding or something. There was plenty of food and everyone welcomed me in warmly, though no one would say what the meeting was about. A man stood up and said, "Now we all know why we're here. I'd like to acknowledge that we have some visitors." Everyone in the room stared at me.

"Go on, admit it," he continued.

"Admit what," I asked?

"That you're an alcoholic."

"But I'm not an alcoholic!"

"You're among friends. There are no cops here. You wouldn't have found out about this place if you did not need our help."

"No really," I exclaimed, "I just wandered in to see what was going on."

Everyone in the room glared angrily at me. Some even clenched fists. Feeling outnumbered about 200 to one, I shouted, "Okay, I admit it. I'm an alcoholic!"

"Hooray," they yelled as they patted my back and stuffed me with more food. It beats dumpster–diving. The leader explained they had to be careful because the police sometimes raid AA meetings to arrest those continuing to drive without licenses. The media also likes to infiltrate the Alcoholics Anonymous to see if any dignitaries or celebrities show up. Many members would lose their jobs if they were found out.

I find discrimination against drinkers particularly galling because of the number of Drunks who developed their alcoholism while serving their country. I once worked for an obnoxious man who got so drunk he fell asleep on the job. When I asked him about his combat experience in Viet Nam he would only talk vaguely about how, when firing a 20 millimeter cannon, you have to look every one you kill straight in the eye. He was having flashbacks which prevented him from sleeping at night, so he slept on the job. For a security supervisor at the Ford Motor Company, this was a damnable thing to do. He should have been fired! It was hypocritical for us guards to apprehend other employees for coming in drunk when our own boss was even more polluted than they were.

I gave him an alcoholism pamphlet out of our medical department and put it on his desk as he slept. When he awoke he threatened to fire the person who placed it there. The other guards all denied it. Thinking I might be fired, the next night I got our breathalyzer machine out, broke an ampoule of the special chemical it used, and placed it carefully on his desk to make it look like I had done a test on him. The next day he fired an inexperienced guard who did not know how to use the breathalyzer. I was not certified to operate a breathalyzer and I could not force him to blow into it without waking him up, so to get back at him, I faked the test by moving the read–out needle by hand and making a red dot on the circular card at the right place to indicate moderate intoxication, then sent it to his boss.

Having faked such a test myself, I have been skeptical about the ability of the police to conduct such tests fairly. When the politicians mandated drug testing for all truck drivers, my employer, Shaffer Trucking, gave me four drug tests in five months. Given the industry–wide false positive rate of one in every twenty tests and the rate at which I was being tested, I could be 90% certain of being fired for drug use in just two years, if I had continued to work there.

When he returned from treatment, the supervisor at Ford was one of the nicest people I ever worked for. The company made him take a tranquilizer to deal with his psychological war injury. With counseling, he got over his problem and became such a good supervisor that he was promoted to head of security. What a shame it would have been if the police had arrested this decorated veteran for drunk driving and, using drug tests, cost him his job, forcing him to drink even more to deal with even larger problems and cheating his employer out of a valuable employee.

Ironic as it may seem, some people need to drink in order to drive. There are other kinds of psychological injuries other than just combat injuries. If I pointed an unloaded gun at your head, you might have nightmares about it for weeks, even though, since the gun was unloaded, you were never really in any danger. An unloaded gun is nothing but a hunk of metal; but if you believe it is loaded, it would frighten you out of your wits.

People who have been injured in car crashes, or even those who have suffered a near miss, get extremely nervous when they have to drive a car. When I was fourteen, a car hit me when I was riding my bike. After my broken leg healed, I found I could ride just fine. The first time a car passed me I ended up in the ditch. The car did not run me off the road. In fact, I cannot remember how I got there. I had a memory lapse. When I got back on the bike, another car passed and I again ended up in the ditch. It was as though an invisible hand had pushed me off the road. For weeks afterward, I could not share the road with a car without my fears taking over and steering me into the ditch. I have felt nervous around cars ever since.

As a stroke of good fortune, I flunked my high school's driver training course because I was so nervous I dug my fingernails into the steering wheel! If it were not for a second driving course taught by a professional taxi cab driver who understood my problem, I might be a bad driver to this day and I would not have written this book. Hiring a good driving instructor was the best eighty dollars I invested in my life. He saved my life many times. Most high school teachers are bad drivers themselves and are certainly not qualified to teach safe driving. Most accidents can be attributed to poor driver training.

It is perhaps that experience that makes me passionate about safety. Many people drink in order to deal with the life-threatening stress of driving. It can be argued these extremely frightened, nervous people might drive even worse if they didn't drink.

My mother once dated a man who flew a B-17 bomber during World War II. When I was a kid, I had a plastic model of a B-17, so I was intensely interested in what it was like to fly one. At dinner, at first he wouldn't answer, but as he got drunk the alcohol loosened his lips: "It was sort of like driving a truck loaded with dynamite," he said; "the plane was armed with thirteen machine guns; all available space in the nose of the plane was filled with thousands of rounds of highly explosive tracer ammunition and firing pins for bombs; behind my head, the turret guns had even more ammunition; behind that, the bombs were filled with thousands of pounds of cordite; to my right and left, the wings were filled with highly flammable aviation gas; in the tail there were even more guns and more ammunition."

"We took off at night," he continued, "together in formation from a big field without lights, and flew just a few feet apart so as not to become separated from the other planes in the group." When I told him about a public television station that was offering rides on a B-17 restored by the Confederate Air Force in order to raise money, he scoffed that a person would have to be crazy to want to fly in one. When I asked him if he still was a pilot, he said that he tried to fly a Cessna 150 trainer once, but the B-17 had a small cockpit and the Cessna reminded him of flying the bomber. When the instructor closed the door, there was a bang and he had a flashback. He had heart palpitations then broke out in a cold sweat and could not fly the plane.

People who have been injured in car accidents have many of the same symptoms. Society gives car crash victims no choice about reliving their "combat" experiences. After a particularly grueling mission where the nose of the airplane was shot away, killing the bombardier and forcing him to cowardly drop his fully armed bombs on innocent French farmers, this B-17 pilot never had to fly a plane again and spent the rest of the war behind a desk. Even fifty years after the war, thoughts of the children he probably killed drive him to drink. Without mass transit available and government negligence allowing the roads to become too dangerous for bicycles, those who see their own children killed just a few feet away in the back seat have no choice but to continue driving. People don't always collide because they have had too many drinks; they sometimes drink because they've had too many collisions. Drunks are scapegoated.

Whenever I see someone out driving a dual–wheel pickup truck with West Coast mirrors and clearance lights on top, I get the feeling that the person inside is trying to pretend that they are me, a trucker, in the same way that the Colonels of the Confederate Air Force try to relive the thrilling, exciting, titillating experiences of being a B-17 bomber pilot. They ride their girlfriends around, trying to be real men, without knowing how stressful driving a real truck really is.

The first time I drove a truck I was terrified. It was more like driving a bulldozer or a road grader than a highway vehicle. "Boy, this is fast," I exclaimed to the instructor: sixth gear, "Boy, this is fast"; seventh gear, "Whoa, this is fast"; eighth gear, "Gosh, this is fast!"

Everybody wants to be a trucker.

   Two hours later it was no different than driving the family car
— but then driving a car became frightening: "Boy, that guy is
close"; "Whoa, that guy is close"; "Gosh, that guy came too close!"
   Sometimes I feel like aiming a gun at reckless drivers and
warning them to stay back. If I did that to a burglar breaking into
my house, there would be no question I was justified in defending
myself. Reckless drivers are far more dangerous than burglars; so
if I threatened one to stay back when he was ten feet away and he
kept on coming, could I be justified in shooting if he came within
three feet, performing a thrilling passing maneuver? If a burglar
kept coming after I warned him, I would be justified in shooting.
But if I shot a car driver, I would probably be sent to prison.
Burglars are poor, lower class, expendable people. Reckless drivers
are wealthy, upper class, aristocrats who contribute to political
campaigns. Should I be content to let wealthy millionaires play
god, taking my life into their hands? They may consider them-

selves qualified to decide by themselves if my life is worth the risk — but I don't! I believe the Second Amendment gives me the right to defend my self from all threats — foreign and domestic!

Most people who drive pickups haul nothing but air. I once moved the furniture of a restaurant chef in Salt Lake City, Utah and could not get my big van down his icy driveway. I said, "Look, we can do this in one of two ways. I can call for help and charge you a lot extra for the shuttle service (loading goods into a smaller truck), or we can put the furniture into your pickup and I won't charge you any extra."

"Oh no, that would scratch the bed of my truck," he exclaimed!

I suppose one has to be a perfectionist to be a great chef; but the bed of his truck had a glossy wax shine and obviously had never hauled anything heavier than beautiful, bikini-clad waitresses.

"I'll put mover's pads in to protect the bed of the truck," I offered. Trepadaciously, he accepted and when we were done he said he would get some mover's pads of his own so he could haul things in his truck without scratching it. I suppose he only bought the truck because, being a chef, he had to prove to someone that he was really a man. The myth is that only men drive trucks, despite the fact that one out of every five licensed truck drivers is a woman who drives with her husband.

Most people drive pickups to win accidents. The narcoleptic, who totaled so many cars that Ford would not let him park on their lot, said he only drove full–size cars like his Country Squire station wagon because, "If you know you are going to be in an accident, you want to be in something that can survive the crash."

"You mean that if you know you are going to smash into a car full of kids, you want to make sure you hit them as hard as possible," I answered — alluding to the fact that when narcoleptics fall asleep behind the wheel, they usually do not apply the brakes.

"Hey, my life is important, too," he quipped. One of the cars he totaled was a pickup. He carved gravestones as a hobby.

People drive large cars so they can smash into things and get away with it. Driving is enjoyable when you don't have to fear being injured. These people do not care about others. They don't care how many people they kill, so long as they are not killed themselves, and they are willing to pay the cost of poor fuel economy and increased air pollution to enjoy the privilege.

When I visit truckstops, I sometimes meet drivers who say they enjoy driving trucks. One guy even said he would do it for free if someone didn't pay him to do it. When I was a kid I used to think the same thing: all the power surging from the huge Diesel engine; being big enough to push other people around. In reality, trucks are top–heavy, under–powered, have terrible brakes, and they can't maneuver well enough to avoid accidents. If I hit anyone, I will certainly kill them and the courts will consider me guilty until proven innocent. The only way a person could enjoy driving a truck is if they did not care whether they lived or died.

I meet people all the time who say that they want to drive a truck in order to see the country. These are usually city folks who do not even ask if they are qualified — they just assume they are. It seems terribly vain that an unqualified person would risk the lives of others in order to experience the thrill of touring the country. In order to enjoy endangering the lives of others, one cannot allow one's self to care about other people. If you care about the lives of reckless drivers, their reckless antics will be stressful to you. You will experience a variety of stress–related illnesses such as nail-biting, hair–pulling, teeth–grinding, hand tremors, upset stomach, lack of sleep, loss of appetite, marital problems, high blood pressure, and heart attacks. This is why most truckers do not enjoy driving; and why good drivers demand such high salaries.

Business executives experience many of the same symptoms

whenever the stock market goes down. The reason executives experience stress is that they care about their businesses, their employees, and their families. Happy–go–lucky people tend to make poor businessmen. They squander their money with bad investments and incautious decisions. They are also poor drivers. People who do not care about their own families are even less likely to care about others sharing the highway with them. This is why the safest, most successful truckers, tend to be self–employed small businessmen.

Millionaires who own trucking companies have made a concerted effort to recruit employees that enjoy their work. Drivers who do not experience stress have lower medical costs, are willing to work for lower pay, and are less likely to seek other employment. Employees who don't give a damn about their families are less likely to object to long absences from home causing them to neglect their children. Do you want someone who doesn't give a damn about his own children to endanger your life, tailgating your car with a forty–ton vehicle?

Trucking companies hire exactly the wrong people. The personality types who enjoy the power and majesty of driving an eighteen–wheeler are the least likely to care about safety. Requiring people to pay for their training in advance tests their desire to drive, not their ability to drive. Imagine that you went to apply for a job and the only qualification the interviewer was interested in was how much money you were willing to pay to get the job. "I don't care how you come up with the money," the interviewer says, "If you want the job, go rob a liquor store or something!"

Government subsidy of truck driver training isn't much better. If you want the taxpayers to pay for your training, repeat after me: "All millionaires are saints; all politicians are heroes; only millionaires are qualified to sit on the supreme court; only aristocrats are qualified to rule the American People." If you are not politically correct and believe that no one should be subject to someone else's authority, then you are not likely to receive money for your training. The worst aspect of subsidizing some but not others is that, as millionaire trucking company executives come to expect new–hires to be trained rather than training them themselves, only the most ruthless of applicants such as those who brown–nose politicians or commit armed robbery will be hired.

Politicians assume that all collisions are accidental; but the preponderance of evidence shows that reckless driving that results in collisions is intentional. To learn what really causes accidents you need a psychiatrist, not a policeman. How can you tell how an unqualified person got a job driving a truck? Or, how a wealthy teen-ager graduated driver's training. Rather than reconstructing on computers what happened in the last few seconds before collision, courts should try to determine which vehicle first approached within one hundred feet, violating the other vehicle's safety zone, and why?

I was driving east on I-44 through Missouri when a hot shot (converted pickup truck) hauling a load of well pipe drove beside me for several minutes. When I slowed down to let them pass they slowed down too. When I looked into the car, an ugly, fifty–ish red–headed woman was smiling at me, pulling her blouse up. When I laughed, she rearranged the pillows inside the car and with many contortions laid down on her back, removed her panties and began masturbating her shaved twat. Judging by the stretch-marks, she was a grandmother, but her boyfriend was trying to sell that meat!

Just outside of St. Louis, on the return trip a little blue Japanese car hovered beside my drive tires. I slowed down to let it pass, but the guy inside slowed down too. I craned my neck to see what he was doing and I saw him slapping his sausage around. Finally he passed, cutting me off, and drove off onto the shoulder. After I swerved into the passing lane to avoid a collision, he accelerated again and assumed his original position by my drive tires, again slapping his sausage around. Suddenly with a rapid side–to–side jerking, he ran me off the road and sped on his way.

People masturbating while driving can best be compared to the phenomenon seen in teen-agers called scarfing. In scarfing, the teen-ager wraps his neck tightly with a rope or cloth and hangs himself from a doorknob or bedpost. As the boy masturbates, he breathes heavy and the scarf chokes off blood to his brain, knocking him unconscious and heightening orgasmic pleasure. Teen–age boys have been found naked, strangled to death in apparent suicide with their hands fondling their genitals.

It appears that fear or near-death experiences heighten orgasmic pleasure. I have seen many couples, Homo and Heterosexual, performing oral–sexual acts beside my truck. These acts might be more common than we realize; because only truckers have the ability to look down and see what most car–drivers are doing. I once read an article in a Playboy magazine about people who prefer to have sexual intercourse on trains or airplane toilets because the fear of being discovered, the writer said, heightens sexual pleasure. Given the number of out–of–wedlock births caused by young men masturbating themselves on the bodies of young girls in disregard for statutory rape laws, one must wonder how many collisions are caused by this form of self–love performed in violation of reckless driving laws.

By far the most unusual deviants I see are those who talk on the phone while driving beside my truck. Psychologists say that suicide is often a form of revenge for some perceived wrong on the part of a loved one. With life insurance companies including anti-suicide riders in their policies, leaving a suicide note can create problems for the survivors. The cellular telephone is a way of leaving a message without leaving a note. The maniac can call his loved one and say, "I'm driving beside a big truck right now; I'm going to drive under its wheels if you don't do what I want."

Can you tell by looking at a chessboard after a game which move out of the thirty or so moves caused the game to be lost? Driving, like playing chess, requires a strategy. You can't get away with waiting to the last second to react to another motorist's reckless maneuver and then depend on lightning–quick reflexes to get yourself out of a dangerous situation. You have to control the space around you and plan your moves in advance. By the time you're in a situation where reflexes are the only thing that will prevent a collision, you have a good chance of losing control of your vehicle just trying to make an evasive maneuver.

This is why the effect of alcohol intoxication has less of an effect on professional drivers than on motorists. About one in ten truckers is a problem drinker — just as in the population in general; but only one in 20 of our accidents are believed to be alcohol or drug–related. Alcoholic Truckers are unlikely to drink very much before work because professionalism will not allow it; but even if they do, they are unlikely to allow themselves to get into a position where reflexes or reaction times are important.

In chess, you wouldn't put your queen next to your opponent's pawn, would you? The pawn might take the queen. Similarly, a skilled professional wouldn't drive his truck near another vehicle while it's driving, because the other vehicle might take his truck. The way to avoid collisions is by not putting one's self into a position where one can get into a collision in the first place.

While an alcoholic trucker might kill someone, replacing him with an inexperienced truck–driving school graduate most certainly will kill an innocent person. The fatality rate in trucking is one death per million miles. Truckers drive three million miles in a typical career. If an expert driver with more than a million miles without an accident drinks a beer in the morning and another at lunch, chances are that ten years later he will have two million miles without an accident even if he gets so totally plastered every night that he has to have a beer in the morning to deal with his hangover. This is because he has learned how to avoid accidents, no matter how drunk he gets.

Regardless of how expertly sober a professional driver may be, he will still never be able to maneuver a truck the way a car can maneuver. Motorists have no excuse for crashing into things with a car. Cars have extreme control over their direction, speed and position. They can be driven off the road without fear of rolling over. It takes two bad drivers to have a collision — one to aggressively violate the law and another to negligently remain still and allow himself to be hit.

Imagine what it would be like if every time someone lost a game of chess they lost their life; if every time they lost a pawn they lost a finger; and every time they lost a bishop or a knight, they lost their hand; every time they lost a rook, they lost an arm! Driving is a lot like that, because people really are killed and injured. You're not playing with carved wooden pieces that can be stood up again after you've lost them. You're playing with your life.

The highway can be visualized as a kind of checkerboard with safe squares and unsafe squares. As long as you stay out of the unsafe squares in front of, beside, and immediately behind other vehicles, you can be perfectly safe. You wouldn't want to drive next to another person because if an obstacle appears in the road, he might need to change lanes. I once ran a Mercedes off the road because I needed to change lanes and the guy inside was talking on the phone instead of paying attention. Every person is going to act to preserve their own life — even if that means taking someone else's life, so you always have to check first, before you try to pass, to see if there is an obstacle in the other guy's way that he might need to avoid. There is no reason to pass someone unless there are mitigating circumstances, such as going much slower than the traffic and other reckless motorists are endangering your life by recklessly passing you. If the person is going faster than the speed limit or within ten miles per hour (or 15 kph.) of the posted speed at night, it should be a felony to pass at all.

The only way you can understand why accidents occur is not by analyzing the ones that happen, but by understanding the ones that don't. There are many people who say we can learn a lot about accidents by studying them after the fact and maybe we can drive differently to avoid similar collisions in the future; but that's only partially true.

For every collision that occurs, ten are avoided, and perhaps a hundred accident situations which could have resulted in an accident didn't. So, to understand collisions you have to look at the whole group; not just the ones that happened, but the ones that didn't — the reckless driving incidents that could have resulted in an accident — and then determine the difference in behavior that caused the accidents to occur.

Skilled professionals already avoid collisions to the best of their ability. Changing our driving behavior to avoid unavoidable crashes would result in ten times as many collisions which are presently avoided. Truckers often face the dilemma of sacrificing one motorist in order to save ten. If we don't, we may need to sacrifice ten in order to save ourselves. Remember, every person can be counted on to protect his or her own life first, no matter what the cost.

In order to minimize accidents in the industry, truck driving school students must learn to recognize reckless behavior long before it results in a collision. If anyone doubts that reckless drivers can be spotted in advance of an accident, I usually remind them that my video camera takes 3–5 seconds to power up, adjust exposure, and focus. I would not have been able to film any reckless drivers if I had not noticed them in advance.

Whenever you see a car–driver drive near a big truck, you should always assume that the motorist operating it is perhaps mentally unstable. Most truckers who break the speed limit do so because they have just–in–time inventory shipments that have to be there. Workers can be laid off if the raw materials they need to produce goods does not arrive on time. There's a big difference between a trucker who's hauling a load that is late and a car driver who feels he has to be on time to his appointment because he doesn't have his priorities straight.

Typically, a car–driver who is late for work is more interested in pleasing his boss than he is in safety. With telephones and faxes, physical presence at business meetings is not necessary. They might be under so much stress that they have a secret death wish. After all, why would they be endangering their life in order to meet some sort of time schedule? A safe driver always takes as much time as necessary to get there safely.

Just–in–time inventories are definitely a cause of recklessness in the trucking industry. Germany and Japan have industrial corridors only 300 miles long linked with high–speed rail systems while American industry is spread out along coasts 3,000 miles apart — linked only by highways. Truckers are routinely and illegally pressured to compensate for this national shortcoming. Their customers may suffer losses if they don't drive recklessly.

I was once paid a $1,000 per day to deliver a computer power supply from San Jose, California to Orlando, Florida. After 24 hours without sleep and soaked with rain, the president of the software company I was delivering to, a defense contractor, came out to shake my hand personally. When I asked why he was willing to pay $1,000 per day to have that thing, he said, "because I'm losing ten thousand per day by not having it." I thought it was a pretty good reason. It is possible to design trucks and busses capable of 100 miles per hour; but as long as the government allows just anyone to drive on the highways, it will never be safe. The car companies disagree. They imply that if you buy their particular brand of automobile, high–speed driving can be performed safely by anyone. Some cars are advertised to be stable at 125 miles per hour or more, and the companies sponsor stock car races on television to prove what their cars are capable of. The stock cars are actually hand–built racing machines with one–piece fiberglass bodies, sharing no parts in common with ordinary cars.

Car companies are selling cars as enjoyable toys to be played with. Car companies try to include reckless driving in their television commercials because they can sell more cars with more accidents. Every time a car is totaled, the car companies sell another car. People would not buy as many cars if it weren't for car accidents. Cars would be built like sailing yachts, with fiberglass bodies and bronze or stainless steel components designed to last forever. Body panels would be painted or replaced to keep the car looking up–to–date and stylish. Their Diesel engines would get excellent fuel economy and produce little pollution. No children would be burned to death in crashes — ever!

Society pays a huge price because politicians who receive campaign donations from auto companies tolerate the sport of reckless driving. Increased air pollution due to poor fuel economy as well as excessive manufacturing results from our culture of cheap, throw–away automobiles with planned obsolescence sold to people who expect accidents. A third of all the insurance sold in this country protects against perils that should not exist.

Suppose guns were advertised on television. Manufacturers could use the same exciting music heard on television commercials; add a few exiting scenes just like in westerns, and people would all want to buy guns instead of cars. Instead of comparing the smoothness of their sheet metal to find out who is a better person, city folks could compare the shininess of their guns. Instead of buying bigger cars, rich people would buy bigger, flashier guns to prove their superiority.

The reason guns, tobacco, and alcohol are not advertised on television is that if they were, every one would want them and they'd likely abuse them the same way they abuse their cars. Instead of joy–riding around just for kicks, people might shoot bullets up and down the street. It is already happening in many city neighborhoods.

Cars shouldn't be allowed to be advertised on television because they encourage reckless driving. Highway workers such as truckers, construction workers, and even police officers have a right to safety in the work–place. Television commercials that show fast, irresponsible driving encourage a kind of recklessness that endangers our lives. We must not allow our highways to be endangered just so that a few big auto companies can stay in business and avoid laying-off autoworkers.

With television stars, and even the police driving recklessly, how can anyone be safe? Car companies like to portray their cars driving fast, without any traffic in idyllic surroundings. The reality is that car commercials encourage so many people to drive when they don't need to that the highways are overcrowded — and this creates hazardous situations for everyone.

Jackknife accident: One must allow for double the usual stopping distance in wet weather. It isn't that the stopping distance is so much greater; it is just that maintaining control while braking is more difficult. All drivers should have skid pad training as part of their driving test to prepare for an accident situation like this.

The Mechanics Of Safe Driving

The Mechanics Of Safe Driving

In traffic, you should always choose the lane that allows the maximum possible stopping distance. There are situations when reckless drivers foolishly try to pass, making a lane change maneuver impossible. I can't explain in words the feeling that goes through my mind when I see a car disappear in front of my front bumper. Motorists commit suicide in front of me and I save their lives. In situations like this, you have to be aware of what your vehicle can and cannot do.

It is possible to calculate stopping distance mathematically using a graph if you know approximately how much time it takes to deaccelerate from a certain speed to zero. Let's say you're going 60 mph. How long does it take for your car or truck to stop? The obvious way to test this is to look at your watch and slam on your brakes, then count how many seconds it takes to get stopped. Lets pretend you are driving an eighteen–wheeler and it takes six seconds to stop at 60 mph. That works out to 10 mph of slowing down for each second you apply the brakes. At 10 mph.–per-second, every second you put on the truck's brakes will slow you down by 10 mph. By knowing how much time it takes to deaccelerate from certain speed to a complete stop, you can calculate very precisely what the stopping distance is going to be at any speed.

First, let us calculate what the stopping distance is at 10 mph. Ten mph is equal to about 14 feet per second. There are 5,280 feet in a mile and 3,600 seconds in an hour, so to convert miles per hour to feet per second you just multiply your mph by 5,280 and divide by 3,600. You will go 14 feet in one second at 10 miles per hour. Since you're slowing down from 10 mph and ending up at zero, you will only average 5 mph or 7 ft. per second while you have your brakes on. How many feet do you travel in 1 second at 7 feet per second? Obviously, you travel 7 feet, so your stopping distance is 7 feet.

To calculate what your stopping distance is going to be from 20 mph, you first have to calculate the distance you are going to travel as you slow from 20 down to ten. Your average speed is going to be 15 or 22 feet per second, so the total stopping distance is 22 feet plus the seven feet when you slow from 10 down to zero or 29 feet.

With air brakes, that is not all there is to it. You also have to take into account the air brake lag. It takes about a half–second for air brakes to activate because pressure must build in the lines before the actuators will start to move. At 10 mph or 14 feet per second, you will travel 7 feet in a half–second so the stopping distance from the time you step on the brake pedal will be 7 plus 7 or 14 feet. That is 7 feet of air brake lag and 7 feet of stopping distance added together. At 20 mph, the total distance will be 22 + 7 + 15 or 44 feet, including 15 ft. of air brake lag.

What you can immediately see from this is that if you double your speed, you increase your stopping distance considerably. Fourteen feet to 44 feet is more than three times the distance. So if

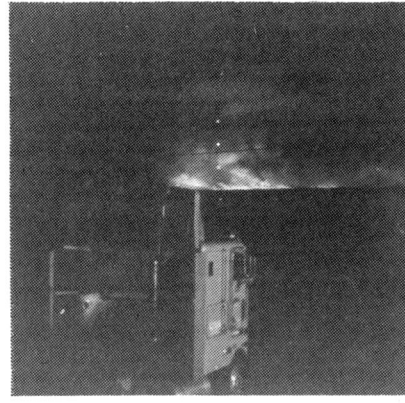

Bad weather and brush fires are just some of the hazards that truckers face.

Triple trailer trucks now compete with railroad trains on many turnpikes and western highways.

This 40–wheeler is designed to spread the weight of the cargo over a long distance to help reduce the strain on bridges.

155

you are going 10 mph and you increase your speed to 20, your stopping distance is going to be three times what it was at 10 mph.

You can continue calculating and find out what your stopping distance will be at 30 mph. At 30 mph you will add 38 feet, so your stopping distance will be 7 + 22 + 38, which works out to 88 feet when you include 22 feet of air brake lag. If you increase your speed from 20 to 30, you double your stopping distance.

Looking at these calculations its easy to see why we have speed limits. If you increase your speed from 10 mph to 20 mph, you triple your stopping distance. If you increase your speed from 20 mph to 30 mph, your stopping distance is six times greater.

This 20–foot tall 72–wheeler is designed to spread it's immense weight without cracking the pavement.

It cannot legally cross many bridges.

156

TRUCK STOPPING DISTANCE
@ 10 MPH. PER SECOND

Speed	Stopping Dist.	Safe Following Dist.	Seconds
10	14	44	3.0
20	44	100	3.5
30	88	175	4.0
40	146	263	4.5
50	220	365	5.0
60	306	482	5.5
70	409	614	6.0
80	525	780	6.5
90	657	921	7.0
100	803	1,096	7.5

CAR STOPPING DISTANCE
@ 20 MPH. PER SECOND

Speed	Stopping Dist.	Safe Following Dist.	Seconds
20	14	72	2.5
40	58	174	3.0
60	130	305	3.5
80	230	464	4.0
100	360	653	4.5

This table shows stopping distances at even higher speeds. It also shows the minimum safe following distance. A vehicle following another needs more room to come to a stop because it takes about two seconds longer for the person following to get his foot off of the accelerator and onto the brake, even if the two vehicles have identical stopping distances. The safe following distance is equal to the regular stopping distance plus the distance the vehicle travels in the two seconds it takes for the driver to react and move his foot at each given speed. Notice that a trucker doing a hundred

157

miles an hour requires nearly a quarter-mile to get stopped and a car requires an eighth of a mile. Following another vehicle at a distance less than this would not be safe because if an object were to fall off the vehicle in front, you would not be able to stop in time to avoid colliding with it.

The "seconds" column is the number of seconds it takes to cover the distance in the safe following distance column at each given speed. This makes it easy to measure following distance while you are driving. For instance, it takes 3 seconds to go 44 feet at 10 mph and it takes 5.5 seconds to go 482 feet at 60 mph. Measuring safe following distance in seconds is convenient because you can easily count how much time elapses between the time the vehicle in front passes a landmark such as a road sign until you pass the same landmark yourself. Notice that the number of seconds needed to be safe increases with speed from three seconds for cars at 40 mph to four seconds at 80 mph.

Politicians and political appointees have killed a lot of people through the years by telling them that two seconds following distance is enough. By packing cars more densely on the highways than is safe, they can delay needed highway improvements in capacity and use dedicated highway tax revenues for other purposes.

If you commute in Los Angeles or any other large city, you have probably seen cars spaced one every 30 feet — ten times the highway's maximum safe capacity. Though metering systems have been tried; none has succeeded in reducing traffic by a factor of ten. Only strict enforcement of reckless driving laws to rid the highways of incompetent drivers and liberal interpretation of civil negligence laws to stop unnecessary joy–riding can do that. People must be educated that self–indulgent shopping trips are not worth risking a life for.

Note that the table does not make a distinction between single

This 60–wheeler is 200 feet long!

Police detained these expert drivers for more than a

half hour at a Wyoming inspections station. During that time, more than 100 drunk drivers drove past without being harassed.

This tag tractor has an automotive–like seat and small steering wheel. The driver said it rode like a diving board every time he drove over a bump.

and multi–lane highways. The safe following distance is 305 feet at 60 miles per hour regardless how many lanes a highway may have or what lane the other traffic may be using. This is because if an object falls off a long flatbed eighteen–wheeler, it is more likely to fall to the side than to the rear.

I was being passed by another truck while going up a hill near Eugene Oregon when one of the truck's tarps fell on the road in front of me. I was only doing about 45 so I had no difficulty swerving into the passing lane to get around it. Suddenly, a stupid idiot doing about twenty miles per hour faster than me whipped around on my right and cut in front of me, slamming on his brakes, stopping in the middle of the road. To save his life, I drove over the tarp. I would have gladly killed him if I'd known what damage it would have done.

I was launched into the air with a bang. My head hit the ceiling. Every loose item in the cab, pens, papers, books, coins, nuts & bolts, tools, maps, my camera, film, and food flew everywhere and wedged under the pedals. The tarp tore off my crossover line and fuel started gushing out from both tanks. I was able to stop one line from leaking by lifting it off the ground and securing it with a tarp strap. The other I had to plug with a pencil wrapped in toilet paper.

Cars came whizzing by without slowing down, kicking the spilled fuel into the air. Some unbelievably stupid people turned their windshield wipers on, even though it was a bright sunny day, spreading the little drops into a thin film across their windshield so they couldn't see where they were going. They could see that the sun was out and it wasn't raining. A group of these idiots were parked about a quarter mile up the road, wiping their windshields with rags. I decided to move because I thought they might cause a collision and the spilled Diesel led right to my truck. Cops have a blame–the–victim mentality. Before the driver could return to pick up his tarp, a guy in a pickup stopped and stole it — right in front of me, despite the fact that the company name was written boldly all over it! I was too busy plugging the leaks to stop him. I followed the driver back to his shop in Eugene where he fixed me up. Who needs crooked cops when there are honest drivers? He admitted the accident was his fault and insisted on repairing my truck. I doubt many city drivers would have owned up to it. Though I cannot remember his name, his act of decency will be immortalized in this book forever.

There was a wreck in Dallas where an object fell off a truck

This 58–wheeler is just half the truck!

This houseboat requires four 18–wheeler size parking places.

CREATION OF A MINORITY GROUP

into the path of about 100 cars who were all driving within 20 feet of one another. Two people were killed, with 15 injured, and 25 cars were damaged. Who's fault is that? Motorists know that loose cargo is hauled by trucks and they know there are pot-holes, and they know that when flatbed trucks go over pot-holes, things sometimes vibrate loose. Are flatbedders at fault because the machinery they haul is assembled with defective fasteners made in China, or if drunk or lazy minimum wage workers don't tighten the multi-millionaire's bolts properly?

Government officials say you only need two seconds following distance because the vehicle in front of you is moving and you only need two seconds of reaction time to avoid hitting it. The real danger is not hitting the car in front of you, but rather that if an obstacle appeared in front of the car you are following, you wouldn't be able to see it with the other vehicle in the way. The car might go around the obstacle and you would drive straight into it.

If tailgating is so reckless and dangerous, why do we see so many truckers tailgating cars? Surely experienced professionals should know better. Most of them are truck driving school graduates who, not knowing any better, drive their 40–ton trucks exactly the same way they used to drive their cars — recklessly. But there are some legitimate reasons for a truck to tailgate a car. We can understand the reasons for trucks tailgating by using stopping distance calculations.

Looking at the table on page 157, a truck following a car needs 5.5 seconds following distance and requires 482 feet to stop from 60 mph. A car requires 130 feet to stop from 60 mph. The difference, 352 feet, will be traveled in four seconds at 60 mph, which is 88 feet per second. This four seconds is the absolute minimum distance necessary to prevent a collision when following a car. At this distance the truck will just kiss the car's rear bumper and no one will get hurt.

162

Government officials included a trick multiple–choice question on many of the Commercial Driver's License tests which experienced drivers were required to pass in order to get their license renewed. It asked, "What is the minimum safe following distance for a truck 40–feet long?" Nearly every driver in the country said six seconds when the manual said four and the expert drivers found their correct answers marked wrong. Amazingly, the manual said that stopping distance is determined by the length of the vehicle: i.e. a truck 60–feet long requires six seconds; an equally heavy 20–foot long truck requires two seconds. Rear–end collisions became so common among the graduates that many fleets stopped replacing front bumpers and motorists became terrified of "monster trucks."

Stopping distance is actually determined by tire pressure and to a lesser extent, type of rubber, asphalt, concrete, or tread design, not the weight or length of the truck. All trucks with 100 psi. tire pressure will have about the same stopping distance regardless of their length. A 20–foot long truck will require the same stopping distance as a 100–foot long truck — about six seconds. Cut-throat competition has forced most companies, including my own, to use smooth "ribbed" tires (like cars have) rather than the safer "lug" design in order to get better fuel economy, and tire manufacturers have been forced to alter carbon content for longer wear at the expense of tread adhesion. So truck stopping distance now is worse than it ever was.

Since these collisions were deliberately caused by government bureaucrats meddling with economic conditions and training materials to increase the density of traffic on existing highways, I believe there should be criminal penalties for officials who cause this kind of preventable death and injury. There is no dispute about these mathematical equations. Any truck driving school student can calculate what experienced professionals know intuitively. There is no argument. There is no room for opinion. The laws of physics are enforced by God.

The only way the government can ignore these numbers is by appointing into office unqualified individuals such as a former mayor to be the Secretary of Transportation. What does a mayor know about driving a truck? Politicians nominate their personal friends into office on the somewhat racist assumption that those most like themselves are the most intelligent and therefore the most knowledgeable candidates available. The problem is that if a genius thinks he knows more about how to drive a truck than the truck drivers who drive them, people die!

Negligent government officials have allowed car drivers to be taught in their high school driver training class that two seconds of following distance is all you need when you cut off a truck. If a car cuts in front of a truck leaving only three seconds (274 feet) of following distance and slams on its brakes, we can tell by simple calculations that the truck will be moving at 35 mph when it hits and it will push the wreckage 88 feet. If the motorist leaves only two seconds following distance, the 40–ton truck will be moving at 50 mph when it hits.

The above table was calculated on the assumption that most cars can slow down at a rate of 20 mph per second, which yields a stopping distance of 130 feet at 60 mph. According to the syndicated program "Motorweek," which tests cars, the best sports cars with anti-lock brakes can stop in 112 to 117 feet, and vans and pickups with higher tire pressure may need as much as 175 ft. When I wrote them that they should test an eighteen-wheeler for comparison, they thanked me for my suggestion and tested a Mack CH 600. Unfortunately, they fudged the test by using the wrong speed — 52 mph for the truck and 60 for the cars. The host John Davis's statement that "178 feet was almost as good as some of the cars we tested" was misleading because, by interpolating using the above table, we can calculate that a truck that requires 178 feet to stop at 50 mph will need about 250 feet to stop from 60 mph. This is better than the football field most trucks require; but more than double the distance of the antilock–equipped car tested on the same program.

It is strange that a nationally–syndicated educational Public Television program would make a mistake that is intuitively obvious to every high school drop–out who drives a truck. Certainly, the producers are just as able to go through the math as I am. Perhaps because the program is about car driving, there is a bias. The producers may desire to make cars and trucks seem safer than they really are.

The down-side of this mathematical distortion is that a very negative picture is painted of truckers. If Motorweek's viewers really believe that an average truck can stop in 178 feet (the test was performed in ideal conditions), they may drive more recklessly than they otherwise would and when the wreckage of their cars (possibly full of children) is pushed exactly 304 feet (truck stopping distance of 482 feet at 60 mph. minus 178 feet), they might blame the trucker for failing to stop, when in order to do so he would have had to violate the laws of physics. Even worse, the program might be quoted as an authoritative source in publications like this book and be used to formulate government policy or, in a courtroom, to wrongly convict someone. Elitism degenerates to bigotry.

What do you suppose happens if a truck tailgates 15 feet from a car and the car slams on its brakes? If both vehicles are doing 60 miles per hour, or 88 feet per second and the car slows down at a rate of 20 mph each second it has it's brakes on, the car will travel at an average speed of 73 feet per second for the first second, which, if you add the fifteen feet in between the truck and the car, equals 88 feet — the exact distance the truck will travel in one second at 60 miles per hour. (In the first second, the trucker will not have time to even put his brakes on.) The collision occurs exactly one second after the car puts its brakes on; the truck will be doing 60; the car will be doing 40; the impact speed will be exactly 20 mph. That means the collision will be less severe if the truck tailgates 15 feet from the car than if it maintains a two second following distance. A 50 mile per hour rear–end collision with a truck will certainly be fatal. The next time you cut off a truck, pray the driver accelerates enough to save your life; because if he did not tailgate you and you put on your brakes, you would certainly be killed.

It is possible to calculate impact speed at various following distances. At 4 seconds following distance, the impact speed will be zero and if the truck is already touching the car's rear bumper, the impact speed will be zero and the maximum force of collision occurs at about one and a half seconds of following distance. This means that if a car cuts in front of a truck, it's safer for the truck to change lanes than to slow down. If there is another vehicle engaged in a reckless passing maneuver, preventing changing of lanes, it is best for the truck to tailgate the car until it can make an evasive maneuver. Increasing the distance between vehicles will increase the severity of the impact if a collision should occur. Tailgating might also intimidate the car driver into driving safely and not put on his brakes, thus avoiding any collision at all.

This does not always work. Once in Washington D.C. a car cut me off, forcing me to change lanes, running two other cars who were trying to pass me off into the median. I blasted him with my air horns when he first veered out his lane and again after I ran the cars off the road. Amazingly, after being honked at twice, he again cut me off and slammed his brakes on, forcing me into the paved median myself, in front of the cars I had run off the road. I continued to drive for a half mile to put as much distance between him and me as possible. Judging by the quality of his Mercedes, he was a wealthy bureaucrat who never did an honest day's work in his life; extremely intelligent, but in chess: mated in seven moves. He was probably frustrated with silly anger at being honked at while stuck in a traffic jam caused by thousands of other foolish people like himself, who stupidly drive one to a car rather than speeding along unimpeded in a bus.

If I had hit him, I would not have argued or exchanged information. I would have gotten out my tire gauge and splattered his head on the pavement — not because he nearly cost the lives of two innocent people, but because I had to risk my life to save his — twice. And because the government would not have prosecuted him. What he did was perfectly legal. I was the one who broke the law. I ran two cars off the road and I drove onto the median to save his life.

In many states it is illegal for trucks to change lanes to make an evasive maneuver except for passing. That's ironic because it's passing which causes collisions. Morally bankrupt elected officials generate revenue from the trucking industry because experienced professionals disobey the law rather than injuring someone, or endangering their own lives. Discriminatory laws have created a public relations problem for the trucking industry because experienced professionals are labeled as criminals while people who aren't competent to drive the cars they own are needlessly risking their lives. Then they complain in the media about "monster trucks" terrorizing them. Am I a terrorist just because I drive a truck for a living? Even "Truckers News" has articles about car-drivers being traumatized by truckers.

Lets read how this type of trauma occurs: The article is titled: "4–WHEELER TRAUMATIZED BY TRUCKERS" by Francine Marcoux of Marstons Mills Massachusetts. (Aug. '93, p. 12)

"The 18-wheeler was in the passing lane and seemed to lose speed as he also made the ascent. For a few minutes we were driving parallel to one another, but he gradually fell behind."

In other words, she passed him on the right and she was driving in his blind spot for several minutes.

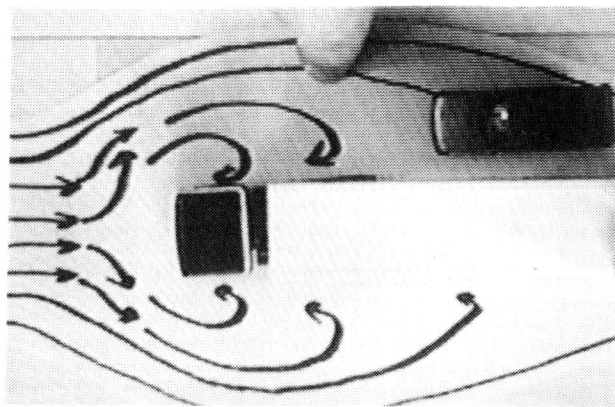

These lines show how the air flows around a tractor–trailer in a head–wind and in a cross–wind.

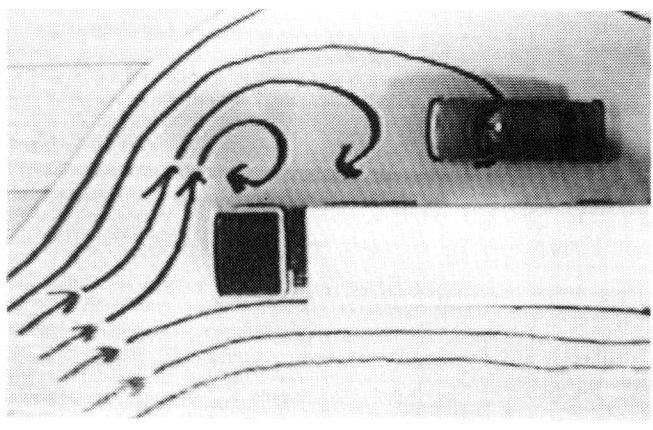

168

"Almost instantaneously, my Honda was struck on the left rear and began a clockwise spin which resulted in another hit at the right rear. The impact propelled my car across the left lane and into the guard–rail and up across the snow bank."

The problem with this story is that if the truck was moving faster than the car and struck the car on the left rear corner to impart a clockwise rotation, there is no way that the car could have been struck on the right rear corner and then moved to the left. The truck was in the way. The car would have ended up on the right side of the truck, not the left. Things move to the right when they are struck on the left — at least that's the way it is in bowling and billiards. The only way this accident could have happened the way the writer said it did is if the car, after passing on the right, had moved aggressively to the left in front of the truck even before the collision took place. Then the two collisions which imparted a clockwise rotation would not have prevented the car from ending up on the left as the writer said. The car's own momentum would have carried it that way.

Why would a writer for *Trucker's News* deliberately cut in front of a truck after driving in its blind spot for several minutes? If we could visualize the slipstream of air around the truck, we could see how this accident might have occurred. The slipstream is formed by the air being deflected from the front of the truck and being diverted around the truck. Turbulent eddies of air form on the sides of truck because of suction behind and beside the truck in a cushion of air that aerodynamicists call the "boundary layer." This ten–foot thick layer of air that is moving at the same speed as the truck lowers the car's wind resistance because the car doesn't face a head-wind when it moves through this cushion of air.

When the car gets near the front of the truck, dangerous things happen. First, there is a very strong blast of air off to the side that is caused by the air hitting the front of the truck and being deflected sideways. This blast tends to push the car sideways. But some reckless drivers, rather than passing quickly through it, counteract this force by steering toward the truck in order to keep going straight; so they have a side–ways force of air pushing their car to the side, away from the truck, and the steering wheels of the car aiming their car toward the truck.

As the car moves in front of the truck, two things happen: First, the steering wheels take over control of the car's direction because the sideways blast of air is no longer hitting the car, so the car cuts in front of the truck abruptly; second, once the car is out of the truck's slipstream it experiences a head–wind, so in addition to the car cutting in front of the truck; it also slows down. The result is that the truck hits the car.

I wrote Steve Sturgess, Editor of *Truckers News*, suggesting that the facts presented were a physical impossibility. Steve answered: "I agree that the physics of the situation are a little confusing; nevertheless the lady is obviously convinced of the circumstances of the accident." An editorial by Roger King, Director of *Communications for the National Association of Truck Stop Operators*, made much of the fact that the truck driver did not stop to render aid. Out of deference for *Truckers News*, I declined to mention in my video what I feel is another likely explanation as to how the accident occurred: that damage to the car was caused in the collision with the guardrail, and that the truck never made contact with the car. The buffeting winds around a tractor–trailer near a mountain top may have been violent enough to make the writer, in her light–weight Honda, think she had been involved in a collision.

It is possible that the driver was illiterate and unable to obtain a Commercial Drivers License. He may have feared persecution by the police. Local cops are controlled by local politicians, who are in turn responsible to the public. They often do what the public wants them to do rather than what they know is right. Trying to convince a hysterical woman who is not competent to drive the car she owns that an unseen wind alone can cause a car to lose control on an icy road would be as difficult as getting her to believe in witches, banshees and demons. Harassing the highly skilled professional for a situation beyond his control might help the police chief get votes. Whatever the driver's reasons for failing to render aid, he was not at fault; the writer was. Passing on the right is considered reckless driving in all fifty states.

In a cross–wind, the wind effect is especially pronounced. Instead of the head–wind being deflected by the front of the truck, there is a wind coming from the side that combines with the head-wind that hits the truck on it's front corner. The front corner of the truck divides the wind into two streams: one along the side of the truck; and another across the front of the truck so the sideways blast of air in front of the truck is especially intense.

If their drivers are not paying attention, cars approaching from behind the truck in any wind may suddenly accelerate and slam into the rear of the truck when they enter the truck's slip-stream. I was heading westbound on I–78 against a strong west wind near Phillipsburg, New Jersey, when I heard skidding and saw in my mirror a car passing me, going sideways. According to the witness, a safety director at a major corporation, the drunken woman had been cruising at about ten miles per hour over the speed limit and suddenly accelerated to about 85 in the 55 zone as she approached the rear of my truck. A few feet before hitting, she cut sharply to the left, crossed three lanes of traffic, and drove into the median. Then, cutting sharply to the right, spun out of control on dry pavement, passed me while going sideways, and hit the side of my trailer tires with the front of her car. If she had merely removed her foot from the gas when she entered my slipstream, the collision would not have occurred. Cars do not need as much power to maintain speed in a truck's slipstream as they do when fighting a head-wind.

If there is a cross–wind, a car entering the slipstream on the side of the truck may accelerate and slam into the car in front of it. Most of these collisions are caused by the car–drivers themselves and most could be prevented if they would only drive at the same speed as the professionals. Unfortunately, in many states there are different speed limits for cars and trucks, forcing cars to pass through these dangerous slipstreams in order to avoid being passed themselves by other reckless drivers — institutionalizing the violence!

Different speed limits for trucks and cars increases the number of tailgating accidents when cars are tailgating trucks. A car following another vehicle will need a somewhat longer stopping distance than normal because of the driver's two second reaction time. If a truck was going 50 mph, it would need 220 ft. to stop. How much distance would a car need if the car was following 50 feet behind, going 60? Looking at the following distance table on page 57, you can see that the car needs 305 feet, so it will plaster into the back of the truck because it needs some 35 more feet than is available. A truck going 60 mph, would need 306 feet to stop. Since cars need 305 feet, the stopping distance would be the same for both vehicles and the collision would be avoided if they were both traveling at the same speed.

If a mass murderer were to kill thousands of people with a gun or a knife, they would get several consecutive life prison sentences, if not the death penalty. How many consecutive prison sentences should politicians get for killing thousands of people with these revenue–raising, death–defying split–speed limit laws? There is no argument about cold, hard mathematical calculations. There is no room for interpretation or opinion about the mathematics of physics. The politicians know with indisputable evidence that split–speed limits cost lives. The theory behind split speed limits is that if traffic is moving safely at sixty-five miles per hour, and everyone is driving safely, keeping five hundred feet apart, the police can pick out the out–of–state truckers and give them tickets for going ten miles per hour over the limit. They know professionals will not obey a law that endangers their own and other people's lives.

Suppose a governor running for re–election made the following stump speech:

> "My fellow citizens, if I am re-elected I will eliminate the state income tax and sales tax. You will get all the services you are used to receiving and these services will be paid for with a hidden tax on out–of–state truckers that pass through our state. Of course, the truckers will demand higher wages to pay for the tax and since most trucks carry food, the taxes will result in higher food prices, but this is a way to make people on welfare pay taxes for a change! Taxing their food will make them eat less and the money can be used to build parks in wealthy neighborhoods and improve highways to make driving luxury cars easier."

Would such a politician be re-elected? Since the danger of split–speed limits is so mathematically obvious and since truckers are the only professionals expert enough to render a scientific opinion on the subject, legislators who enact these laws fall into two categories: negligent man–slaughterers who failed to seek out educated opinion prior to enacting the law; and first degree murderers who listened to the experts and deliberately chose to ignore them in order to generate revenue. A grand jury should be convened in each state to hear the evidence and punish the wealthy bureaucrats who interfere with ordinary Americans to endanger their lives as they go about their work.

In order to drive as safely as possible, it is best to keep a running tally of cars passing you while you are driving. Add one for every car that passes you. Subtract one for every car that you pass. Try to make the number zero so that the same number of cars are passing you as you are passing. This way, you will minimize the number of reckless drivers who drive their cars and trucks too close to you.

This is sometimes hard for me to do because my truck doesn't go faster than the speed limit on most highways. Most cars go faster than I do. For economic reasons, it would cost me a whole lot of money, blowing fuel out the stack, if I were to drive flat–out all the time. My truck is geared for 57 mph and that's the speed I prefer to go.

Sometimes I have to share the road with government subsidized truck–driving school graduates with fast trucks who haven't been taught that they should slow down when visibility deteriorates. These drivers are paid by the mile and their income is so low, they feel they must endanger other people's lives in order to earn a decent living for their families. They are told they must throw on the iron and maintain their delivery schedule regardless of conditions.

I once roomed with a man who was leaving the United Parcel Service in order to work for North American Van lines.

"Aren't you giving up a lot of pay to come to work here?" I asked.

"There are times when your life is worth more than money," he said.

"What do you mean?" I asked.

"The first time I jacked (jack–knife accident), I swore I would never drive on ice again, but they laid this story on me — how a Black Man will have a hard time finding work and that I better deliver on time."

"The second time I jacked I said, 'this is it,' and started looking. They can get rid of those doubles and triples (two or three trailers) and get some real trucks if they want to drive in bad weather."

You can't share the road with graduates safely. You have two conflicting goals: first, to drive at the same speed as the traffic, and second, not to drive too fast for conditions. In wet or slick conditions you will need about double the stopping distance that you would normally need in dry conditions, or about 11 seconds of following distance at 60 mph. If visibility is less than 11 seconds, you'll have to slow down. If traffic is moving too fast to slow down, then the only safe thing to do will be to park it and tell your boss that you will be late because there are too many unsafe drivers on the road. Someday truckers will shut down the highways to protest unsafe working conditions.

The biggest problem with maintaining the speed of other traffic has nothing to do with the weather. It has to do with the fact that trucks don't have adequate horsepower to maintain speed on hills. As a result, cars tend to approach from behind and sometimes even pass the truck as it reaches the crest of the hill. As the truck starts down the other side the opposite problem occurs. Trucks don't have adequate heat capacity in their brakes to use the brakes continuously while going down hill. If the driver does not release his brakes after a certain period of time, his brakes will overheat and he will lose them completely. Organic friction material gets so hot that it sublimates into gas and this gas forms a kind of air cushion between the friction material and the metal drum. A cushion of air is the best lubricant there is. Astronauts use air cushions to simulate zero gravity. Brakes that are "on fire," or sublimating have no friction at all and produce a tell-tale plume of smoke with a distinctive smell as terrible to truckers as the smell of death.

Trucks descend mountain grades slowly, in low gear so the truck will not speed up out of control when the brakes are "fanned" (released temporarily to cool them off). In order to keep the brakes as cool as possible in preparation for descending grades, experienced truckers allow their speed to increase down shorter hills rather than using their brakes. This creates problems with cars cutting off trucks when merging onto the highway because they do not realize how quickly gravity can accelerate a 40–ton truck down a hill.

The problem with truck brakes is their shape. Metal expands when heated and when the metal in brake shoes expands, it changes the curve of the brake shoe so that the brake pads do not fit the curve of the drum precisely. This wasn't a problem with asbestos pads because asbestos is an insulator and protected the brake shoe

Super Single tires are needed to fit disc brakes on a truck. Even with drum brakes, they provide better brake ventilation. They are not yet legal in all

states because of a myth that they cause pavement damage.

from heat. But modern semi–metallic pads conduct heat.

When the brake drum expands, a gap forms between the brake shoe and the drum. This is why the Department of Transportation is so concerned about properly adjusted brakes, because the brake shoe has to be pushed farther in order to contact the drum when heated. But when the shoe itself expands, cracks form in the friction material caused by rivets stretching the material, and the shoe no longer fits the drum precisely. It touches only at one particular spot on the end, which quickly gets hot and loses its friction.

If you're curious why vocational vehicles like busses and dump trucks don't lose their brakes, its because the end that touches the drum wears out so that the shoe eventually fits in the hot position, and then they have trouble with howling and squealing when they cool off. Truck brakes will work well hot or cold, but not both.

The only configuration that's safe for modern semi–metallic brake linings is disc brakes. The advantage of disc brakes is that they are flat rather than curved, and move in and out in a straight line, rather than pivoting. If they expand considerably in size, they will still fit the flat sides of the disc perfectly.

The problem with fitting disc brakes on trucks is that the large rotors and calipers will not fit inside dual–wheel assemblies so single tires would have to be used. The discs would have to be nearly as big as the tires themselves and the tires would need to be made larger to accommodate them. Asbestos brake dust causes lung disease and cancer, so no one favors going back to asbestos brakes. Based on the crack–pot assumption that single tires cause pavement damage, politicians have made disc brakes illegal. The basic principle of democracy is that 99 fools can make a better decision than one expert and the basic principle of republican government is that wise leaders listen to the experts rather than lobbyists. If the fools see it in their interest to ban safe brakes because they know that trucks with unsafe brakes have no choice but to

Diesel-electric trucks of the future will be shorter, lower, wider, and safer than today's trucks. Decreased frontal area will

make them more fuel–efficient and they will be capable of speeds over 100 miles per hour.

break the speed limit and generate revenue for the government, they will ban the new technology and deliberately allow innocent people to die in order to obtain revenue.

The reason truck tires cannot be made taller to accommodate the larger discs is that trucks would be even more top-heavy than they already are. Although trucks appear to have tires eight feet apart, the springs are only three and a half feet apart, so there's a lack of initial stability. Trucks sway back and forth quite a bit. If the swaying becomes too extreme, the momentum can cause a truck to roll over — the leading cause of death among truckers.

Sway also causes pavement damage as the weight shifts from one side to the other. This overloads the wheels on one side of the vehicle and then on the other. Axles ten feet wide would weigh a lot more than axles eight feet wide; but they could carry significantly more weight. If a set of double trailers eight feet wide has axles loaded to 20 thousand pounds and sways in such a way that 15 thousand pounds of that moves to one side then the other, a ten–foot wide straight truck that did not sway could carry 30 thousand pounds on each axle without doing any additional damage to the road.

If politicians could be replaced by honest citizens, trucks would be a lot safer. We would get rid of the idiotic eight foot width restriction; and once trucks are ten feet wide, the cargoes can sit

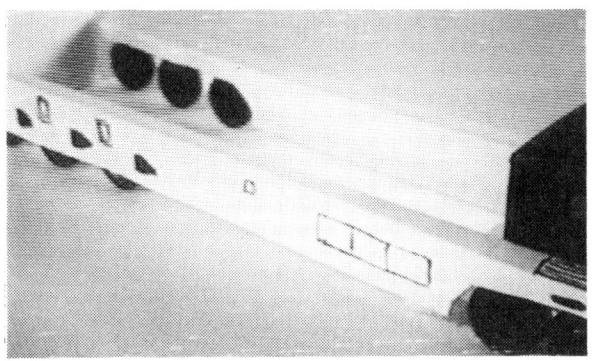

The electric motors will be mounted out-board of the tires. To change a flat tire, the shiping container will have to be dropped on the ground.

down in—between the wheels instead of on top of them — that way they won't have the tendency to sway and to rock all over the place in strong winds the way they do today. No vehicle should be required by law to be taller than it is wide. The active–hydraulic suspensions will kneel to allow the truck to drop its cargo container directly on the ground, thus eliminating the excessive time drivers spend away from their families waiting for their trucks to be unloaded.

The trucks of the future will be smaller and shorter than todays trucks. In fact, they will be no taller than the average motorhome. Bridges can be made lower, with shorter spans, less expensively. Since the mirrors on trucks are already ten feet apart, having the wheels ten feet apart as well would not increase the number of accidents — only the severity of the accidents.

Accommodating ten foot wide trucks on existing two lane roads will involve compromises. For one thing, there will need to be two center dividing lines instead of just one. The truck will need to

Trucks are required by law to be built so light that refrigerated trailers are made of aluminum and fiberglass covered foam. Forklifts can poke holes in

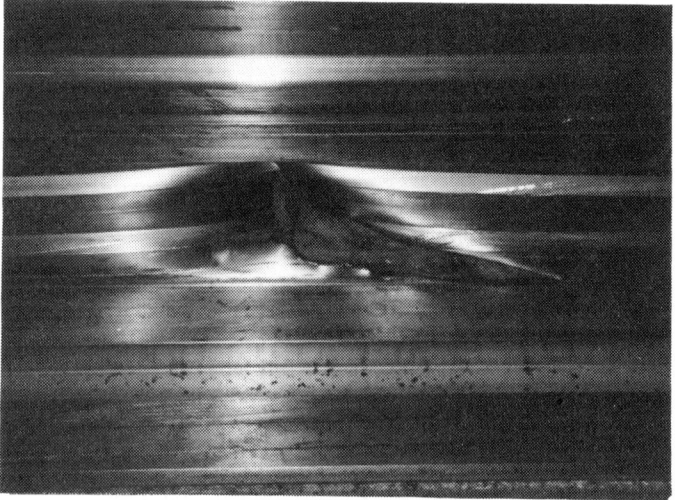

the inside walls, exposing the foam, and trapping bacteria which cannot be washed out.

take a little bit more room as it drives down the road, and cars will be restricted to a narrower lane on the side. There will also be a bicycle lane that the trucks will not drive on because super–single tires will crack the asphalt if they are routinely driven too close to the edge. Trucks will only drive there if two trucks need to pass each other.

This arrangement will have beneficial side effects because, if cars are restricted to narrower lanes, they will be kept farther apart and there will be fewer head–on collisions. Also, the police will be able to spot incompetent drivers more easily because weaving in and out of the lane will be much more obvious if the lane was narrower.

Driving beside trucks will have to be discouraged. There will need to be a $100 fine for driving within 100 feet of a truck and a $1,000 fine for unnecessary passing. On a four–lane road off the highway, the truck will need to straddle both lanes to prevent traffic from passing, just as today's trucks do when making a wide right turn.

Diesel–electric propulsion has been on trains for over fifty years. We would have had it on trucks 20 years ago if it was legal.

This truck was required by law to be built so flimsy that the wheels fell off.

The advantage of diesel–electric power is that engine horsepower can be augmented with power from batteries to give the truck more power for climbing hills, so it can maintain the speed of the other traffic. When going down the hill, the electric motors can be thrown into reverse to provide a braking action and store the energy in the batteries to be used over again. Because of this, the fuel economy of the truck will be significantly better — between 10 and 15 mpg. The drawback with diesel–electric is that the weight of the batteries alone would get the truck thousands of dollars in overweight fines per week. The government policy of discouraging the advance of technology by limiting the weight of the truck rather than the weight of the cargo has far-reaching consequences.

My truck is used to haul raw meat. Then, it turns around with a load of ready–to–eat candy and ice–cream. The holes in the floor of my trailer (see photo) contain bacteria that contaminate the candy and the ice–cream. The damage is caused by inexperienced, minimum–wage fork–lift truck drivers. But, the damage would not occur if the government would allow trailer manufacturers to build trailers heavy and strong enough to resist the damage. The politicians say that weight restrictions are necessary to protect the roads, but Michigan trains (with their 11 axles and 42 tires) weigh considerably more than a typical 18–wheeler, and they don't damage the roads. The extra wheels and brakes and lighter axle loading actually makes them safer.

The reason the government limits the weight of the truck rather than the weight of the cargo is that politicians take huge campaign donations from food warehouses and other retailers that require lumpers to unload the trucks. If this new technology became available, many of those political patrons would go out of business.

Suppose the half–acre or larger supermarket you buy food at now, with it's 40 employees, had to compete with a supermarket no larger than a gas station and only five employees — with food loaded at factories onto shipping containers that plugged into the walls that you could just walk into to buy things. This supermarket could be located near your house and you could get there by bicycle or electric golf cart. You wouldn't have to drive a car.

Inside these containers you would find products arrayed on shelves, with an aisle down the center, just like in a supermarket. This will save millions of dollars of board feet of timber each year by not having to use pallets and corrugated paper boxes in order to ship things. The active-hydraulic suspension on these lower, wider trucks of the future will be so smooth that it will not be necessary to use additional packaging materials in order to protect the cargo.

Shopping malls in the future will consist of a central aisle as they do now, but instead of stores, shipping containers loaded at factories will be plugged into the walls and the goods will be arranged inside exactly the way the manufacturer will like them to be displayed. The mall owner will get a salesman's commission of ten percent on each item he sold; but you will be buying the goods directly from the manufacturer so there will be no mark up. Everything you buy will be wholesale plus ten percent. The cash register will be at the doors, just as in any department store.

There will be no janitorial staff. A single janitor will drive a floor–sweeping machine around the building at night. The shiny white ceilings, floors, and walls of the containers will be hosed out once they are empty and sent to a different manufacturer to be loaded with other goods. When a new container arrives, everything left in the old one will go on sale. Goods that the manufacturer wants to push will be loaded onto movable racks in the central aisle of the container and moved out into the mall to be displayed. The containers will be rented by the manufacturer from a leasing firm and will be shipped by rail to a major city, where a self–employed trucker will pick them up to make the final delivery.

The wide, low, Diesel–electric container trucks will be capable of 100 mph. One hundred mph busses will replace cars. Why would anyone want to drive at 55 mph when they can ride comfortably at 100 mph?

Decisions about how wide, or tall, or heavy a truck should be should be made by the expert who has to drive it, not by corrupt politicians bought by special interest campaign donations. Government officials know this technology exists because it was the government who developed it for improved military logistics. By not allowing civilians to use military technology, politicians have deliberately and knowingly destroyed forests and endangered species by harvesting trees to make expendable paper boxes. They have caused cancer and acid rain by destroying the ozone layer with car and aircraft exhaust, increasing dependence on foreign oil by forcing people to drive cars and fly planes rather than riding fast, fuel efficient busses. They have caused drunk driving and other death and injury by forcing incompetent drivers to continue to drive even when they don't want to. They have endangered the health of all Americans by making healthy sports like walking, running, and bicycling too dangerous to participate in; created delinquency in the cities because streets are too dangerous for children to play in.

My first truck. The author bought this White–Freightliner for just $4,900 and rebuilt it for another $7,000. It's winey, turbocharged and supercharged V-6 bus engine leaked oil all the time and was not up to the task pf pulling a set of doubles. It made a lot of money pulling furniture and electronics, though.

Conclusion

Conclusion

A ncient Greeks believed that democracies should be governed by citizen–representatives drawn by lot rather than elected politicians. They reasoned that elected government would become more of a popularity contest than an expression of public will. If a state legislature were appointed into office by drawing names from a barrel demographically representative of all registered voters, would this group of 53% Women, 12% Black, 9% Hispanic, 63% lower middle class vote to ban technology that reduces food prices by 30%, oil imports by 20%, and fatalities by thousands of lives per year with safer, more efficient vehicle design? Recognizing that reckless driving is a predominantly an upper middle class White Male phenomenon, would these women and minorities allow just any macho man to drive a car without first proving himself competent in an intensive and sophisticated training course taught by experienced professionals? Would they let anyone drive a 40–ton truck with only a two–week training course?

The first real truck I drove was a fire truck at the Ford Motor

Company. I was a security guard-fireman. I later became a security courier, responsible for delivering payroll checks, computer disc drives, gold watches, trophies, and award plaques. Although the van I drove was no larger than most people's personal vans, the half–ton of computer printouts I carried made me a trucker, of sorts, and that was enough to get my foot in the door at North American Van lines.

At my first interview, nearly all of the 120 other applicants with me were government–subsidized truck–driving school graduates. Seeing I was the least qualified person there, I almost walked out, thinking I had no chance of being hired. Then our manager, Jerry Garbe got up and said, "We're going to be doing drug testing, so if you can't pass the test, you might was well leave now." A third of the graduates got up and walked out. I was astonished. Our elected representatives had used my tax money to train drug users to drive trucks!

Then Jerry said: "We're going to check your driving record, so if you've had an accident or a ticket you might as well leave now." The crowd hushed, then murmured. A guy stood up and said, "I've had an accident. Who hasn't — if they've got any experience? How am I supposed to pay back my student loan if I can't get a job?" "You've already proven that you can't even drive your own personal car safely," Jerry answered "Why would you expect us to hire you to drive a truck? If you had experience delivering pizzas and you did that safely, we would hire you without truck driving school."

Of the four of us who were hired, none had been to school, and all had prior experience. One failed his drug test, but found another job with a fly–by–night company that received a Job Training Partnership Act grant for training him even though he was already experienced. I am proud to say I worked my way up in the industry and did not take shortcuts. I did not have to pay for my training. But the Job Training Partnership act paid twice, first when I learned to drive at North American, then all over again when I changed jobs and went to work at Mayflower. Mayflower double dipped. Even the Burtons, North American Van Lines' award–winning Electronic & Trade Show drivers of the year, had to attend basic truck driver training and the National Safety Council defensive driving course just so Mayflower could get a subsidy.

The training subsidies were so large that many trucking companies adopted a new management style, preferring to hire inexperienced, subsidized drivers rather than paying the higher wages needed to attract qualified ones. Of the 14 that graduated with me at Mayflower (other than the Burtons) three had accidents their first day. One, legally drunk, ran into a low bridge. Another hit a shag tractor (small single–axle for spotting trailers) while backing out of the company shop. The third got stuck trying to turn around in a farmer's field near our motel because he thought tractor–trailer tires loaded to 6,000 lbs. each were big enough to drive in soft loamy mud without sinking in. The innovative progenitor of this management style, J.B. Hunt, shrewdly kept the rates low so that his major customers like Wal–Mart would bail him out when the time came to pay all the jury awards he owed to those injured and killed by his inexperienced drivers. J.B. refused to hire the many drivers I met who were laid off from competing companies due to this low–balling because experienced drivers were not eligible for government subsidy. It would have been better if he had defrauded the government, sending experienced drivers through training programs like Mayflower's rather than violating equal opportunity laws that require employers to hire the best qualified applicants.

To be fair to Mr. Hunt, The J.B. Hunt company has indeed shown dramatic improvement since congress cut off the Job Training Partnership Act money — mainly by hiring the best qualified

CREATION OF A MINORITY GROUP

applicants and paying them a living wage just as the companies he ran out of business had done. Many of his customers who initially welcomed the lower rates discovered too late that instead of several excellent trucking companies competing for their business, they became locked into an expensive low–quality oligopolist without any real competitors.

With much fanfare, J.B. published big quarter–page ads proclaiming that he had 20% fewer accidents in '93 than he did in '92. Does that mean he killed and injured 20% more people in '92 whose lives would have been saved if he had merely chosen to change his management policy sooner? Shouldn't he be charged with negligent manslaughter for deliberately injuring all those additional people when an almost trivial change in company policy would have prevented 20% of the collisions? If I seem unfair to J.B. Hunt in singling him out for doing what a hundred other companies have also done, it is because he threatened to sue me for including a scene in my first video, "So You Want To Drive A Truck?," of one of his drivers making a delivery with two flat tires on one wheel. The driver had phoned his dispatcher from a truck-stop to tell her he would be late for his delivery while he got his tires fixed. The woman told him to air them up and make the delivery on time. They were almost empty when he arrived and one was completely flat when I filmed it. Tire mechanics call tires that have been run flat "zippers" because many have been killed when they exploded with a characteristic "zipper" appearance along the edges of the fragments. Threatening lawsuits to silence whistle-blowers is a nasty way of doing business — and unpatriotic.

According to Bennett Marine Video, who publishes my sailing video, "How to Equip a Trailer–Sailer for Serious Ocean Cruising," there are more than 200 videos available on the subject of sailboats and sailing. Most of these are professionally produced, including some 20 about boating in the nude. Isn't it strange that

The author's present truck is a Peterbilt 362 with a 310 hp. Caterpillar 3406 driving a straight nine–speed and 3.55 rears.

He pulled the "world's largest box of M& Ms" while working for the Mars Candy Company

The author clowns with a friend during routine maintenance.

on the subject of trucking, there are only two safety videos available to truckers; both produced by me, the most successful of which was made with equipment no more sophisticated than a home movie camera, a VCR, a TV, and a table lamp with the lampshade removed?

Although many companies have produced videos for in-house use, the threat of lawsuits prevents them from being sold. My competitors must charge $50 to $100 dollars per tape (beyond the ability of most drivers to afford) to pay for their legal defenses. They must be politically correct. They cannot, for instance, film a 20–foot dump truck following a car with only a two second following distance, as recommended in the *Uniform Model Commercial Motor Vehicle Handbook* (published by the Department of Transportation), because the car would be destroyed if it applied it's brakes. The threat of lawsuits requires them to film an instructor standing behind a podium lecturing students to do what the handbook says, rendering their products boring, incomplete, and without relevance to the real world.

My own videos are criticized for being needlessly opinionated. (Don Long, *Inside Trucking*, Feb. '93, p.9) Dissatisfied corporate customers write that they cannot use my materials because they do not contain "facts." I make them this way to protect me from lawsuits. Opinion is protected speech. Factual representations are not. Even if a producer filmed proof that the Federal Handbook was wrong, he or she could still be sued because courts consider government training materials to be authoritative. Representing fact as opinion allows me to avoid lawsuits. Showing actual footage of reckless drivers in action lets students see for themselves that the handbook is wrong. This leaves students confused, but an experienced professional instructor will almost always tell the truth to students — even at the risk of losing their job. I have not yet received a single letter from any company representative who suggested any opinion rendered in any of my videos was wrong. The Department of Transportation is changing the handbook to match my opinions. I have not changed any of my opinions.

A student who has doubts is more likely to ask an expert for advice than one who thinks he knows it all. The problem with any kind of diploma is that it gives the recipient the idea that he or she knows all there is to know. I have personally seen a graduate tell a million–mile perfect driver how to drive better. Although it was obvious from the graduate's speech that even driving a car was completely beyond the young man's ability, the expert was willing to listen to him because, as he put it, "You have to know what's in the minds of these guys so you can figure out what they're going to do." Car drivers drive recklessly for the thrills. Truck drivers drive recklessly mainly due to incompetence or indecision.

The statistical probability of an average driver achieving a

The author's first video, filmed with a home video camera.

195

million miles without an accident based on the industry standard of one crash per 100 thousand miles is one in a thousand. Since most companies seem to have one million–mile expert for every thousand drivers, it seems their million–mile perfect drivers are statistically luckier, not safer. The best drivers are understandably self–employed in order to make more money. This leaves them out of company safety statistics. While I have read of a fleet of 100 trucks having 13 expert drivers with 25 million perfect miles between them, the safest companies of all are partnerships between a handful of expert drivers who have never had an accident.

Before no–fault insurance laws, self–employed experts had a huge competitive advantage over big companies because they did not need to buy liability coverage. Today, the government requires experts to pay more for insurance than companies pay for new–hire graduates. Such huge cash reserves are required for self–insurance, that only the largest and ironically the most dangerous companies can save money this way. Drivers are not the only ones who fear liability. Many driving schools are afraid to use textbooks to train drivers because teaching materials such as my own often disagree with the Uniform Model Commercial Motor Vehicle Driver's Handbook. If a student has an accident as a result of such a difference of opinion, the school might get sued, even if the text-book is right; because the courts consider government documents to be authoritative.

The problem with any book on driving is that no one accident avoidance strategy works in all instances. What works in most situations may cause disaster in an unusual situation. This is not a problem for experienced drivers who can see unusual traffic problems developing; but when training incompetent people, text-books can never be comprehensive enough to include every possible contingency. Just as it is impossible to train a novice to be a chess master in two weeks, it is impossible to train an inexperienced

conclusion

A highway sign encouraging reckless driving.

A truckload of bombs.

Signs like this should be mass-produced and planted on every highway entrance ramp

197

driver to become qualified to drive a tractor–trailer in two weeks. Schools are forced to use the government handbook because only the handbook claims to be able to do this.

In the opinions of many of the expert drivers I interviewed, the *Handbook* kills more people than it saves. Expert drivers are already preventing collisions in the majority of accident situations they get into. By advising students to try to prevent the non-preventable collisions that make up the bulk of Federal accident statistics, the Handbook encourages evasive maneuvers that are inappropriate to avoid the much greater number of collisions now being avoided. Accidents which are successfully avoided do not show up in the statistics.

By monopolizing the textbook business, the government prevents students from studying authoritative material written by experts. The *Handbook* cannot be considered authoritative because the name(s?) of the author(s?) is (are) not mentioned. Who wrote the Uniform Model Commercial Motor Vehicle Driver's Handbook? Expert drivers say it seems to have been written for the trucks of ten years ago, when the government official(s?) who wrote it last drove professionally. It seems to be concerned more about obeying laws made by incompetent regulators and corrupt politicians than honest safety and professionalism.

I once bought a book from the government called *"The American Practical Navigator,"* written by Nathaniel Bowditch, who died in 1838. Though published in 1802, it mysteriously includes chapters on radar and satellite navigation. It is now in it's 70th edition. The fact the book is still in print suggests that in the year 2183, truck drivers might still be learning safe driving from a manual written for the trucks of the 1960s and '70s. Textbook writers like myself and publishers would like a chance to compete with the government for this business.

Exactly why the *Driver's Handbook* contains errors is a matter of some debate. It seems there may be some hard–core racists in the Department of Transportation. In the past, the way one became a trucker was to work one's way up from the bottom. One would start out as a lumper, loading and unloading trucks like my dad when he was a teenager. My dad was nearly killed when a truck axle he was unloading fell on his chest, stopping his breathing. In a feat of superhuman strength, he and two others rolled the thousand–pound axle over his head, giving him a cauliflower ear for years afterward. His boss became so impressed with the care with which he handled the cargo and his attitude toward safety, he was invited to drive a fork lift and to spot trailers in the parking lot. He drove a straight truck in the army for two years before he was ever invited to drive an eighteen–wheeler. Drivers in the past learned to drive where their accidents could be supervised before being allowed to drive where their brushes with death were not supervised.

In the 1950s, most lumpers were White; so it is no surprise that in the '70s most truckers were White since truckers started out as lumpers. Today most lumpers are Black. Based on historical demographics, we should be able to predict that most non–rural truck drivers today should be Black. That hasn't happened. In creating a short cut for unemployed suburban Whites to become truck drivers, truck driving schools have created a glass ceiling for up–and–coming Blacks. After Black employees had proven themselves qualified under intense supervision, corporate executives were influenced with taxpayer's money to pass them up for promotion and hire unqualified Whites whom the government had approved for student loans. Why would a businessman train a Black employee at his own expense when the federal government will subsidize him to hire a White employee already trained through the Job Training Partnership Act?

The reason for this overt discrimination is the public's reaction to Black truckers. I once was hiring a Black man to help load some furniture when the Mayflower agent he worked for said, "Don't hire him, the customer is Morman." (Movers like Mormans because they have huge quantities of food and building materials in their basements that they keep in preparation for nuclear war. They pay so much to have this stuff shipped that moving companies are willing to do almost anything to keep them as satisfied customers.) The White men the agent had picked out for me did not seem as competent as the Black man. He told me we had a full blown grand piano along for the ride, so I felt I needed the best men I could get, and reluctantly hired the Black against his advice.

Not long into the move, the customer accused my Black man of stealing a gold bracelet from on top of his wife's dresser and insisted that I fire him immediately. "Did you see him steal it," I asked?

"No, but he was up there," he answered."My White guys were up there too," I said; "How do you know they didn't steal it? If I fire all of them there will be no one left to move the furniture."

When we arrived in California, I had to hire another Black man to unload because there were no qualified Whites. This new man then threatened to quit because the customer called him a "nigger." I had to do something to smooth the situation, so I ordered the man to assemble the piano. I thought if he could impress the customer with his skills, the customer might like him. When the piano was done I sat down, cracked my knuckles, and played "Crown Him With Many Crowns," a favorite warm–up of the Morman Tabernacle Choir, to prove that pianos do not go out of tune just because they've been dragged across the country 3,000 miles and reassembled by a Black Man. The customer was not impressed. "I don't mind if you bring a nigger into the house I'm leaving," he shouted, "but I sure as hell don't want one getting his 'ape paws' all over the house I'm moving into!" Even grocery ware-

200

houses sometimes express displeasure with a Black man handling boxes of food with ungloved hands.

I ordered the man to tear down the load, the job I normally do, in order to keep him away from the customer. My White guys then threatened to quit because they felt I was giving the Black man preferential treatment. A minor problem occurred when he had to go to the bathroom. I told him: "Don't ask, just do it!" But, then I followed him in because I thought he might be mad enough to whip out his meat and pee all over the house. Fortunately, he was too professional for that. Finally, we found the dresser the bracelet had disappeared from and I had him bring it into the house while still wrapped. I assumed an abnormally relaxed posture on the piano bench and ordered him to remove the pads in front of the customer and remove the gold bracelet from the top drawer, left side, and give it to the customer. When he said, "You mean this bracelet," the customer's jaw dropped open! "You knew it was there all the time," he bellowed! "No I didn't," I answered; "I knew the man was a well–paid professional and that he had no reason to steal."

Before the government liberalized weight laws so that sleepers could be attached to trucks, drivers had to stay at motels and flophouses. Because of discrimination, Blacks had to to sleep across the driver and passenger seats. You can imagine the safety problems caused by not getting adequate sleep.

Because of these extreme types of customer relations problems, industry executives began talking about a "driver shortage." What they really meant is that there is a shortage of White Males willing to abandon their families for long periods of time. There is no shortage of Blacks. The greatest scourge of the Black community is that the welfare system will not allow unemployed men to remain with their families while on welfare.

These issues do not receive public exposure because most of the trucking print media has been sold to outside interests. Trucking magazines no longer maintain truckers on their staffs. There is no longer any independent free press available to speak out against governmental abuses. The few magazines still run by former truckers face stiff competition from a number of free tabloids. The largest of these, *Truckers News*, is actually owned by the National Association of Truckstop Operators (NATSO). Its name is misleading because it is not published by truckers (note no apostrophe in the title), nor is what it publishes actually "news." Winning several awards against other tabloids such as the *"Star"* and *"National Enquirer,"* it's front page headlines are often exaggerated, if not deliberately false. One headline, designed to increase fuel sales at NATSO member truckstops charging as much as 10¢ more per gallon, accused non–NATSO truckstops of selling contaminated fuel that poisoned people and destroyed engines. (Sept. '93, front page headline) *Landline*, the publication of the Owner/ Operator/ Independent–Driver Association (OOIDA), the editor of which is a former trucker, correctly reported that there was no proven evidence of any individual having been poisoned or any engines destroyed — though it reported that one driver's engine was destroyed when he bought fuel contaminated with gasoline at a NATSO member truckstop. (Oct. '93, p.26)

Clearly, a conflict of interest exists when a truckstop chain calls itself a newspaper and distributes propaganda for free in competition with legitimate magazines sold in the same truckstop. In the lawsuit ACLU vs. Detroit Edison, the American Civil Liberties Union forced a power company to stop distributing free incandescent light bulbs by successfully arguing that Edison was monopolizing the light bulb market in violation of federal trade laws. Almost overnight, people began replacing their power hungry incandescent light fixtures with more energy–efficient flourescent lighting. The suit significantly lowered electric bills and atmospheric pollution in the Detroit area.

The free press is our single most important pedestal of democracy. Citizens cannot vote correctly if they do not have correct information on which to base their decisions. The blatantly illegal policy of providing free or below-cost advertising space to advertisers in a free publication when they agree not to advertise in competing publications sold for several dollars each, monopolizes the constitutional concept of a free press in the same way as the power company monopolized light bulbs. *"Landline"* can only be obtained by buying a subscription from the Owner-Operator/ Independent Driver Association. It is not available in NATSO truckstops. *"Truckers News"* is distributed free.

One time I saw a driver throwing a huge pile of *Truckers News* into a dumpster. When I asked him why, he said the issue contained a misleading article favoring a company he had just quit.

The author performs routine maintenance — checking the power steering level. The large cylinder above the exhaust pipe is the air cleaner.

The company, he said, was dangerous and he did not want other drivers to make the same mistake he did of going to work for it. Tabloids often exaggerate the good qualities of competing companies in order to make them seem more attractive to each other's drivers. By encouraging drivers to change jobs more often, the tabloids can sell more want ads.

There may appear to be a broad variety of informational publications distributed free to truckers, but most of these are published by only a few oligopolistic conglomerates like Chilton's or Newport Communications. Newport publishes *"Heavy Duty Trucking," "Truck Sales & Leasing," "Truckstop World,"* and *"Phonefacts,"* as well as *"Truckers News."* I quote from the *"1993 Newport Communications Rates & Editorial Calendars,"* page 18:

"Advertisers who select *Heavy Duty Trucking* or *Truckers News* on an exclusive basis will earn additional space credits beyond corporate and frequency discounts. For instance, fleet market advertisers buying four pages in *Heavy Duty Trucking* but who do not advertise in (Chilton's) *Commercial Carrier Journal* or (Intertec's) *Fleet Owner* will earn a fifth page free. Owner/Driver market display advertisers buying four pages in *Truckers News* but who do not advertise in (Randall's) *Overdrive* or (Chilton's) *Owner/Operator* magazines will earn a fifth page free."

Is it any wonder that *Truckers News* has grown to twice the size of it's nearest competitor?

"The price of truth," as *Overdrive* used to call itself, has risen at ten times the rate of inflation. "The voice of the American Trucker," as *Overdrive* called itself when it was first bought out by Randall Publishing, now calls itself the "magazine for the American Trucker." Though the present staff are excellent writers, none has ever driven a truck for a living and the editorial has taken an anti-trucker tone.

The present editor, G.C. Skipper, accused drivers from Colonial Freight, one of the oldest and most-respected trucking companies, of tailgating him while he was riding around in a car. Amazingly, in the same editorial, he admitted his own car was being driven recklessly. In a reply, the president of Colonial questioned if the incident even occured.

Bobby Seale, who publishes *Truckers/USA* says I should not be so hard on Randall Publishing because they saved *Overdrive* from bankruptcy. *Overdrive* was boycotted by its advertisers after it editorially stated that the strikes of '73 and '78 would benefit truckers. *Overdrive* was right. Trucker's incomes rose so much that school teachers joined trade unions in an effort to achieve income equity with us. Truckers became huge supporters of the Democratic Party and helped Jimmy Carter and Bill Clinton get elected. According to Bobby, the strikes were what caused the oppressive regulations in the first place as the government acted in response to the threat of truckers' power.

Bobby refused to print a half page ad of mine supportive of another strike planned for election day 1994 that never materialized. The ad merely described the numerous constitutional violations committed against truckers by state and federal governments and indicated that my company, Trucking Video, supported the strike. He did eventually run the ad, but it was heavily censored and contained a number of typographical errors. It did not sell as many videos as I would have liked.

G.C. Skipper at *Overdrive* once "hit the ceiling" according to his advertising salesman over an ad I submitted titled, "95% of What is Written in Trucking Magazines is Not Written by Truckers." Although the name of the ad was written on the check I sent, he substituted a press release published in *Trucker's Connec-*

tion Magazine without my permission. Later the salesman refused to sell me any more space because a couple of large trucking companies objected to my products. The ad he created out of the press release (amazingly) did okay, so I didn't lose any money. I just didn't make as much as I would have with the original ad.

Bobby explained that the "true customer" of any newspaper or magazine, particularly one that is given away for free, is not the reader, as many suppose, but the advertiser, since a magazine with an editorial policy or ads that offend it's advertisers will not remain in business for long.

My sister, who owns a free rock & roll tabloid distributed in the Detroit area, once published an article about racism in prison written by an inmate in Kalamazoo, Michigan. Four guards strapped him face down on a table and beat him severely about his thighs and buttocks while holding a copy of my sister's magazine in front of his face. Another inmate wrote her what had happened and she published that too. The fact that a senator had to get involved before the problem was resolved proves how rampant censorship has become in the United States. Ironically, the same prison bought one of my videos for it's truck driver training program. The thought of that man loose in an eighteen–wheeler after being treated that way by law enforcement makes me shudder. Her strong anti–government editorial policy has restricted her magazine's growth. Most big retailers will not buy ad space from her.

My videos clearly save lives. But Kent Powell, President of *Truckers News* and *Heavy Duty Trucking* will not let me advertise in his publications — not because he is adamantly against saving lives, or because he is trying to encourage as many accidents as possible in order to sell more trucks; but because I included a segment in my first video, *"So You Want to Drive a Truck?,"* that described how I was defrauded by a NATSO member truckstop.

This hurts, because these free tabloids have more than twice the circulation than all the legitimate trucking magazines combined. The illegality of these trade practices was aptly demonstrated in ACLU vs. Detroit Edison. Free distribution is a form of lowballing. Giving away free ads in return for not advertising elsewhere is more than just illegal, it is unpatriotic, since monopolizing information destroys democracy.

Truckers have been harmed by the lack of a strong free press to take a stand on important issues such as government regulations and police corruption. Law enforcement agencies have become increasingly unregulated, and behave like businesses selling a product — with police officers pressured like salesmen to perform. Because I started as a security guard, I can say I've been on both sides of the badge; but the Ford Motor Company had a more sensible attitude toward law enforcement. Experienced workers are too valuable to be fired over trivial matters like drinking, drug abuse, or petty theft. Ford had rehabilitation programs for that. Rehabilitating workers costs only a tenth as much as training new ones. The Ford Motor Company now out–competes both the Japanese and Europeans internationally. American trucking companies, who see their best workers quit because of police harassment and large fines, are so burdened with lawsuits, they fear competition from Mexicans and Canadians taking our jobs away from us in our own job market!

Over–policing used to hurt the Ford Motor Company. The first plant I worked at, a farm-tractor plant in Romeo Michigan, had more than a million dollars worth of theft annually. The thieves were so well organized, they had engines built with duplicate serial numbers over in England just to be stolen. Because the wealthy executives didn't realize the stolen engine parts were being assembled into complete vehicles, they hired security guards to watch the trash, the scrap-pile, and the gates. The 78 guards never revealed how the thefts were occuring because the average Ford plant has only 24 guards and most of the 78 would have been laid off if the thefts ever stopped. The plant was eventually closed due to inefficiency, but twelve of the guards kept their jobs — guarding the empty building!

If reckless driving was ever stopped, most police officers would be laid off. The American people could not afford a huge paramilitary force of highly skilled and trained people that contibute nothing productive to the economy unless they were self-supporting. In the trucking industry, inspectors write fines for defective equipment so trucking companies are forced to lease, rather than buy, brand–new equipment, rather than repair older trucks at huge cost. Inspectors fabricate violations that don't exist in order to meet their quota.

I was stopped at a scale in North Carolina a few months ago (1993), just two miles short of my destination. The inspector said I was out of hours even though I knew from my odometer that I had not over–driven. The inspectors required me to spend eight hours at the scale without being allowed to sleep. I asked to use the bathroom, but all four of them put their hands on their guns and their leader told me I had better stay in my truck. When I finally was permitted to leave, I was so tired I could not keep between the lines and ran two cars into the ditch. I was driving flat out, as fast as the truck would go, exceeding the speed limit by 15 mph. because I had to go (to the bathroom) real bad. When I got to the customer, I parked in the middle of the narrow driveway and ran inside yelling, "Where's the John?" When I came out fifteen minutes later, they threatened to call my boss because their employees could not get into the parking lot with my big truck in the way. A hundred cars were parked up and down the street!

At the trial, the other criminals all had attorneys, but I had to represent myself after 32 hours without sleep. The prosecutor announced that if I didn't plead guilty, he would delay the trial and that would cost me thousands of dollars. I told him I'd rather pay the fine than miss work any day, so I pled guilty. When I showed the judge the dated, time–stamped supermarket receipt that proved beyond any shadow of doubt I was innocent of driving more than ten hours without stopping, the judge was furious.

"Do you know what 'guilty' means," he asked?

"Yes," I answered. I was so tired, I had failed to show proper respect.

"When you talk to me, you show proper respect. You don't talk to me like I'm your next door neighbor or something," he warned.

"Yes sir, your honor!"

"That's better. Why did you plead guilty if you thought you were innocent?" he asked?

"You just heard the prosecutor say if I didn't, I'd lose thousands of dollars!"

He hissed, but he had to let me off with a stern warning to log personal shopping trips. One of the inspectors, who was moonlighting as a deputy sheriff because his insurance wouldn't pay for his child's kidney problem, apologized to me four times. He explained that inspectors are mostly good people, but they are under a lot of pressure.

The reason corrupt politicians are able to exploit truckers and police officers alike, is that the industry is denied the ability to self–regulate. Any other community of a million people would have their own mayor, town council, and police force. Truckers are spread out over more than 100,000 miles of federal highways, making it impossible to create these institutions. The Teamster's Union used to perform a police function, but its leaders were charged with racketeering.

Imagine if the mayor of a city was charged with racketeering just because a police officer apprehended a criminal a little too violently. The trucking community is like a city, but our laws are not recognized by the government. I once was present when a J.B. Hunt driver was beaten after backing into another truck in a parking lot. He begged for mercy, claiming it was an accident, but the five guys kicking him said it was intentional. They said he knew he wasn't qualified to drive an eighteen–wheeler, but chose to drive it anyway — that was intentional! Any one who is not safe at three miles per hour in a parking lot will not be safe at 55 mph. on the highway.

If the same driver had jack–knifed on an icy road because his tractor had a dangerously short wheelbase, the same five drivers would have rescued him, given him a Thermos of coffee and kept him warm until help arrived. One type of accident violates our laws of professionalism and the other does not. Why the violence? There is no other way of enforcing the laws that we depend on for our own safety.

What would truckers do if we achieved power? The Secretary of Transportation would be elected from a group of candidates with a minimum of one million miles without an accident instead of being appointed by the President from among his personal friends. Police officers would need a minimum of a quarter-million miles without an accident before being hired, and they would patrol with a video camera, two to an unmarked car, without guns. Traffic patrol would be a highly–skilled specialty among the police instead of a place for rookies to pay their dues before moving on to something else. They would send large fines by mail to reckless drivers without the need for high–speed chases. The weight of truck cargo would be limited instead of the weight of safety equipment, so trucks could be built as heavy as necessary to be safe. Logbooks would be required, but the company would get the fine if a driver did not get home to rest once a week. There would be no $1,200 fines for arithmetic errors on logbooks or for forgetting I went to the supermarket. Laws impossible to obey for safety reasons, such as speeding, would be ruled unconstitutional. A one–year apprenticeship as a co–driver would be required to get a license to drive an eighteen–wheeler.

In the trucking industry, so many government policies are based on political expediency rather than scientific research that finding competent experts to oversee the development of training material is impossible. Producers and textbook writers have to make a choice between teaching students to be safe, or teaching them to be legal. A training program cannot be considered authoritative if it arbitrarily takes sides on controversial issues.

This book is, therefore, an autobiography about my personal experiences as a self-employed trucker that contains personal advice about how to drive more safely. While it would have been preferable to interview expert drivers about safety, that would expose me to lawsuits because advice correct in nine out of ten situations can backfire in the one out of ten that is an exception to the rule.

Although I am an experienced survey researcher with a degree in Sociology and Economics from the University of Michigan, well acquainted with scientific methods of opinion research, this is not a product of science. Although I interviewed expert drivers exhaustively prior to writing this book, all opinions are my own.

Demographically, truckers are an extremely diverse and predominantly rural group, though city folks have made inroads thanks to truck driving schools. Most drive because there is no other occupation available. Farming, ranching, and logging all require land, which is in short supply. Most city drivers will go back to local jobs when the economy improves, so trucking is a boom and bust industry. There are six times as many people licensed to drive trucks as there are jobs, but low wages, crack–pot regulations, and intolerable, life–threatening working conditions create a continual shortage of long haul drivers. When wages increase, such as during an economic boom, truckers can make as much per hour as doctors, but most drivers choose to stay home and raise their families — remaining impoverished.

Ruinous competition is defined as a competitive situation in business that results in either the death or injury of the participants, or a criminalogenic work environment where obeying the law and staying in business is impossible, or in an oligopolistic, or non–competitive economic result. In trucking, ruinous competition is measured by the number of families which dissolve, or sons on drugs, or daughters turned promiscuous as a result of long absences from home. These problems are blamed squarely on the government for interfering with our ability to regulate the amount of time a driver may spend on duty without returning home to rest.

The present hours–of–service regulations contain a loophole that allows companies to order drivers to log off duty when they are actually on duty but not working. As a security guard at Ford, 98% of my time was spent doing absolutely nothing. The 2% of the time I actually did any work, I saved the lives of two employees, extinguished seven fires, and detected a potentially very expensive chemical spill. Should I not be paid at least the minimum wage for all my time spent at work, whether I did any real work that day or not? Should not a trucker be paid the minimum wage from the time he arrives at work until the time he returns home? Not including time spent eating or sleeping, a driver that spends seven days on the road works 100 hours per week. The minimum wage of $4.25/ hr. for the first 40 hours and $6.33/ hr. for the next 60 hours plus $30/ day meal expense as estimated by the Internal Revenue Service, is $700.00 per week. If a driver is paid 20¢/ mile and drives 2,000 miles he will earn $400, but his restaurant expenses will be $20/ day or $30/ day before taxes or $210. $400 - $210 = $190 which divided by 100 hours equals $1.90 per hour — substantially less than minimum wage.

Truckers are scrupulously honest. Most business is conducted over the phone. You will rarely see truckers shake hands because a trucker who does not keep his word will soon be drummed out of the business. Truckers are required by law to lie in their logbooks by pretending to be off-duty when they are really on duty. Truckers are the most heavily armed minority group in America, with nearly a third of them carrying guns — two-thirds in some companies. Strikes tend to be violent. If the government requires us to violate our own code of moral conduct, what other moral codes are we likely to violate?

It seems our industry is heading toward another violent strike. Strikes rarely raise the freight rates for more than a few months, but they can be economically effective if the rates are increasing, as the increase can be made to occur suddenly, rather than gradually over time. But this strike is not about money. Drivers have had to drive twice as many miles and spend less than half as much free time at home in the '90s as in the '70s to make the same amount of money. Since the fatality rate per mile has not changed, this makes the job twice as dangerous and the neglect of our children: twice as great.

Technological development in trucking has lagged behind autos because of government interference. Todd Spencer of *Land Line* wrote that my first video was too negative and that many of my remarks would be interpreted as "anti-safety" — an Orwellian word for political correctness. In a way, I can sympathize with Todd's reason for refusing to allow me to advertise in his magazine, better than Ken's at *Truckers News*. I have publicly stated that legislators should be held accountable for crimes of negligence committed while in office. The Owner–Operator Independent-Driver Association must lobby these same corrupt legislators.

The failure of the OOIDA to support the non–violent work stoppage in November '93 suggests that grandfather organizations

The author lubricates a suspension hanger with a compressed air–powered grease gun. Knowing how to grease the truck is a necessary requirement for truckers. It is impossible to know how to inspect a truck for safety defects unless one also knows how to lube it. Students should not be given a license to drive a truck without first proving they can lubricate one.

Do not install any equipment on a truck that can't be adjusted with a sledgehammer. The author gets his hands dirty changing a shift–linkage universal joint. Above his hand are the secondary fuel filter and the ether injection canister. The fuel–water separator is under his elbow.

214

may no longer be useful. The OOIDA is run by men who have not driven a truck in a decade; who have little in common with most drivers today. The destruction of the Teamsters Union was a disservice to drivers and to society because the longer protests are delayed, the more violent they will be. I offer this book in the hope that legislators will be inspired to change their minds before violent men hold them accountable for what they have done.

Many viewers of my films have written that they would like to see a program about what the life of a typical trucker is really like. It is difficult to put into words and even more difficult to film the loneliness, boredom, and sheer terror that characterizes my line of work. So, I wrote a parody of Bruce Springsteen's "Born to Run" song, itself a satire on American life. Many of the words are his and it is sung to the same tune:

> In the day we sweat it out on the street
> loading someone's American dream
> At night we drive down endless highways
> in suicide machines
> Sitting on engines above highway nine
> eighteen wheels, fuel injected
> and keepin'it between the lines
> But baby this truck breaks the bones in yer back
> Its a death trap, a suicide rap
> I've got to get another one
> 'cause tramps like us: baby we're just born to run!
> ********
>
> Jackie(old girlfriend) come with me
> I want you to see where all our money comes from
> Just lay right down in the sleeper berth
> and let this old bitch (truck) bounce you around some

'cause baby we can't break this trap
we've gotta run 'til we crap
baby we can't go back
Oh ride with me for a little while
'cause babe I'm just a tired and lonely driver
and you've got to know how it feels
You got to know love is wild
and I got to know our love is real
****

Between the clouds of turbo-lag smoke
we scream down the boulevard
Lizards comb their hair and wave you near and the gays
try to just look hard
the truckstop lots are parked up full
at the start of a foggy moonless night
we're gonna die together on the road tonight
if we don't get there before light
****

The highway 's jammed with broken heros
on a day late all night run
My bird dog 's barkin' a constant hum
but I'm tryin' to beat the sun
But baby I don't know why they have us
put up with the madness here on this highway
But someday girl, I don't know when
we'll get to that place that we really want to go and we'll
head for home
But 'til then
tramps like us; baby we're just born to run!

216

If you would like more information about trucks or trucking, Bill Trescott's Trucking Videos can be ordered by sending $19.95 per title plus $5 shipping and handling to:

TRUCKING VIDEO
PO Box 4265
Sargent Texas 77404

"SO YOU WANT TO DRIVE A TRUCK"

Produced with a home movie camera; shows the realities of life as a trucker, with emphasis on burdensome government regulations.

"SARGENT TEXAS RECKLESS DRIVING VIDEO"

Demonstrates safe driving strategy by comparing defensive driving to the game of chess. Television commercials are shown which encourage recklessness and safe vehicle designs are introduced which are illegal under state and federal regulations.

"THE SECRETARY OF TRANSPORTATION'S MESSAGE TO TRUCKERS"

Shows Federico Pena's speech to truckers near Austin, Texas. It also includes an interview with a Michigan Police Academy instructor and an introduction by Todd Spencer of the Owner Operator Independant Driver's Association.

"HOW TO SUCCEED AS AN OWNER/OPERATOR"

Shows the equipment needed to run a profitable trucking business. It includes segments on how to do a grease job, taxes, and how to calculate self-employment income.

All titles are approximately one hour in length.